Telecommunications Cost Management

For a listing of recent titles in the *Artech House Telecommunications Library,*
turn to the back of this book.

Telecommunications Cost Management

S. C. Strother

Artech House
Boston • London
www.artechhouse.com

Library of Congress Cataloging-in-Publication Data
Strother, S. C.
      Telecommunications cost management / S. C. Strother.
          p.   cm.—(Artech House telecommunications library)
      Includes bibliographical references and index.
      ISBN 1-58053-178-4 (alk. paper)
      1. Telecommunication—Cost control.   I. Title.   II. Series.

      TK5102.5 .S77   2002
      658.4'038—dc21   2001056653

British Library Cataloguing in Publication Data
Strother, S. C.
    Telecommunications cost management. — (Artech House
    telecommunications library)
    1. Cost accounting   2. Telecommunication—Cost control
    3. Telecommunication systems—Cost control
    I. Title
    658.1'553
    ISBN 1-58053-178-4

Cover design by Gary Ragaglia

© 2002 ARTECH HOUSE, INC.
685 Canton Street
Norwood, MA 02062

International Standard Book Number: 1-58053-178-4
Library of Congress Catalog Card Number: 2001056653

10 9 8 7 6 5 4 3 2 1

Contents

1

Introduction

Every day, we use telecommunications products and services. We take for granted that we can pick up the telephone, dial a few digits, and talk to someone across town, or across the world. We have come to expect quality telecommunications services, and we are surprised, or even offended, when the phone does not work.

The cost of telecommunications is one of the top five expenses listed on the balance sheet for most businesses. But corporate users do not make up the largest slice of the telecommunications pie. About 70% of telecommunications revenue comes from consumer spending. It is not unusual for an individual to spend more than $250 a month on personal telecommunications services. This book is written for corporate users. Small businesses, large corporations, and everyone in between will profit from the information contained in this book.

The telecommunications industry is a unique mix of contrasts. Next-generation technologies such as asynchronous transfer mode (ATM), wireless access protocol (WAP), and digital subscriber line services are taking the market by storm but most phone calls still originate across plain copper wires. Mature carriers such as AT&T compete with upstart carriers such as Qwest for dominance in today's fast-paced marketplace.

Those who cannot keep up with the pace are being left behind. Smaller carriers are being swallowed up by hungry "supercarriers" through a record number of mergers and acquisitions. If customers do not stay abreast of the market, they end up using outdated services and being overcharged. Customers have more service choices than ever before, and carriers are enjoying record revenues. One thing has not changed, however—*few customers understand whether the charges on their phone bills are accurate and if the rates are competitive.*

Every month, more than 10 million American businesses that do not fully understand their phone bills dutifully mail (large) checks to the phone companies. If they do not pay these bills each month, their services will be turned off, and no business can operate without phones.

Most carriers use billing systems designed in the 1970s. These "legacy systems" were designed to handle only a small number of service offerings. They cannot keep up with today's complex telecom services, and they end up being inaccurately billed. Few businesses are assured that they are not overpaying. They do not know exactly how charges are calculated or how to verify that those charges are correct.

Why this book?

This book is designed to help businesses *understand* their telecommunications services and then effectively *manage the costs* of those services. The book explains in nontechnical language the most common telecom technologies and services in today's marketplace. These services fall into four categories: local, long distance, data, and wireless services. Each service offering has its own unique type of bill. Sample phone bills are included in the book. The reader will learn how to read, interpret, and reduce these phone bills.

This is not a technical book; it is written for businesspeople who want to effectively manage their telecom services. Eventually, at some level in a company's organizational chart, a nontechnical businessperson becomes ultimately responsible for the functionality and expense of the organization's telecom services. This book is written for those people.

Cost management: Audit, optimize, negotiate

Each section of this book contains cost management advice that falls into three categories:

1. Phone bill auditing—Verifying charges and detecting errors;

2. Service optimizations—Reducing your cost by selecting different services or different service providers;

3. Negotiation—Successfully negotiating low pricing, avoiding future penalty situations with your carriers, and negotiating credits and refunds.

The negotiation tips are especially valuable because phone companies have a unique corporate structure. Their customer service staff is often isolated from the sales staff that is assigned to manage the customer. This book offers sound advice on interacting with customer service and provides detailed strategies for negotiating favorable rates and services with the carrier's field sales staff.

When customers neglect their telecommunications services, they always end up paying too much. Customer ignorance always results in greater revenues for telephone companies. Most carriers prefer to have a well-informed customer base, but they will not turn away a customer who does not mind paying too much. A business can return significant dollars back to its bottom line by using the three approaches (auditing, optimizing, and negotiating) taught in this book.

How to use this book

A customer's viewpoint of its telecom service is more often *reactive* than *proactive*. The strategies contained in this book can reverse this trend by taking control from suppliers and giving it back to customers. Some businesses aggressively manage their telecommunications expenses, but most have no corporate strategy for managing these costs. Their telecom decisions are dictated by the phone bills that show up in the mailbox each month. Local, long distance, data, cellular, and paging bills get paid each month, and no one ever knows if the bills are accurate. But two things remain true for every business: They want their phones to work, and they want low rates.

This book is organized according to customer's view of telecommunications services. The book's major sections focus on local service, long-distance service, data service, and wireless service. Each section explains the carriers' technologies and service offerings. Simple cost management tips are then given that specifically apply to these services. For example, the reader should first have a basic understanding of dedicated long-distance service, before trying to save money on that service.

A person with significant expertise and experience in the telecom industry will already be familiar with some of these strategies, but may still find fresh ideas in this book. Many of the book's concepts are illustrated with case studies of companies in today's marketplace. Your company might be in a similar situation. A busy person with little time to read may prefer to use this as a reference book. Use the checklist in Chapter 3 to guide your own phone bill audit. Before paying another phone bill, read the chapter that relates to that particular service. You may gain just enough know-how to save a few hundred dollars—or even a few thousand dollars—on that bill.

However you choose to use this book, I am confident it will be a valuable tool in your telecom cost management efforts. This book is a culmination of more than 10 years' research and experience. This experience includes working *for* the carriers as an employee, *working with* the carriers as a consultant, and *competing against* the carriers as an entrepreneur. I hope the ideas contained on these pages will return thousands of dollars to your bottom line and, ultimately, help you regain full control of your telecommunications.

2

Background of the telecom industry

To work most effectively with telecommunications today, it is helpful to be familiar with the industry's background. The telecom industry has been in a constant state of change during the last century and a half. The cycle of growth in this industry has been: invention, deployment, monopoly, and then government regulation. Scientists first invent the technologies and then businesspeople deploy them. Soon, a single company such as Western Union or AT&T develops a monopoly and then the government enacts strict legislation to regulate the industry. This cycle has proven true with the telegraph, the telephone, and the computer.

The initial *invention* phase began in the late 1700s when inventors experimented with sending electrical signals across wires. On May 24, 1844, Samuel F. B. Morse sent the first telegraph message from Washington, D.C., to Baltimore. The telegraph then became a mainstream technology for the next 100 years, used by businesses and private citizens. On March 10, 1876, Alexander Graham Bell's words "Mr. Watson, come here. I want you," traveled across wires, and the age of the telephone was born. Bell went on to found the Bell Telephone Company that soon became one of the world's most powerful corporations.

Since then, a growing network of wires has encircled the globe as customers have eagerly subscribed to telecom services. This *deployment* phase of the industry is highlighted by the first transatlantic undersea cables between the United States and Europe, as well as the claim in AT&T's 1908 annual report [1] that it wanted "a phone in every home." That vision has become reality, and now billions of people use the telephone every day.

Once a new technology such as the automobile, telephone, or computer has been invented and widely adopted by consumers, two things happen. First, a dominant supplier emerges: a company such as AT&T. Shortly thereafter, the government steps in and begins regulating the industry.

In telecommunications, this began to happen around 1910. Telecom technology was rapidly improving through advances such as the advent of wireless service, the use of copper wiring, and the invention of the vacuum tube. At this time, most foreign telephone companies were government-owned and functioned as inefficient monopolies. But it was a different story in America. AT&T capitalized on the new technological advances and aggressively grew its business. AT&T emerged as the dominant American telecom supplier. It even gained control of Western Union, the only other telecom giant. But AT&T's competitors, and the government, attacked AT&T's monopoly.

During the time between World War I and the 1980s, AT&T controlled the entire telecom industry except for the independent telephone companies, which mainly operated in small rural markets. Some of these companies, such as Sprint and MCI, would eventually grow to become serious competitors of AT&T.

From AT&T's first major legal defeat in 1913, when the government ordered it to spin off Western Union, until the 1980s, the company's monopoly was constantly under fire. Finally, in 1984, after 70 years of lawsuits, the breakup of Ma Bell was ordered. AT&T could keep its long-distance business, but it was to spin off the Regional Bell Operating Companies (RBOCs) that would offer local telephone service. The industry had completed its evolution from a free market system to a tightly regulated marketplace.

The Telecom Act of 1996 loosened some government regulation and is often referred to as the "deregulation of the industry." But a phone call today is subject to the laws of no less than three governing bodies: federal, state, and local. Compared to the previous 150 years of telecommunications history, today's marketplace is far from being truly deregulated. Today, up to 30% of a typical phone bill is made up of taxes, surcharges, and miscellaneous fees.

A concise timeline of the telecom industry

The key moments in the development of the telecom industry are highlighted in the following timeline, which serves as a concise history of telecommunications. Many forces have molded the telecom industry, but none more profoundly than technology, economics, and government. These major events make up this brief chronology of telecommunications.

1792   Frenchman Claude Chappe invents a working telegraph.

1844   Using Morse Code, Samuel Morse transmits a telegraph between Baltimore, Maryland, and Washington, D.C., across the first telegraph line in America.

1851   England and France are joined by the first undersea telegraph cable.

1861   The first transcontinental telegraph line is completed. Telegraph lines from the Atlantic and the Pacific are joined at Ft. Bridger, Utah.

1866   Western Union solidifies its position as the dominant telegraph company by acquiring its chief competitor, the United States Telegraph Company.

1866   Government regulation begins with the Post Roads Act, which empowers the postmaster general to fix prices for government telegraphs.

1866   After 8 years of unsuccessful attempts, the first transatlantic undersea telegraph cable is completed. The cable connects the United States to Europe by way of Newfoundland and Scotland.

1876   Alexander Graham Bell, a teacher of deaf students, files a patent application for an "improvement in telegraphy" just 3 hours before Elisha Gray files for a similar patent. Over the next 11 years, Gray and other inventors file more than 600 lawsuits claiming they are the true inventors of the telephone, not Bell. None is successful, and Bell is forever immortalized as the inventor of the telephone. Bell offers the rights to his telephone patent for $100,000 to Western Union, which declines the offer and starts its own telephone company one year later.

1877   The Bell Telephone Company is formed by Alexander Graham Bell and a handful of partners. Coinventor Thomas A. Watson is the company's only employee.

1877   The American Speaking Telephone Company is started by Western Union using Gray's patent.

1878   The Bell company files a lawsuit accusing Western Union's telephone company of infringing on Bell's patent. In a late-night settlement reached in 1879, the Bell company agrees to stay out of the telegraph business and Western Union agrees to stay out of the telephone business. Just 9 years later, Bell violates this agreement when its subsidiary, AT&T, starts to transmit telegraphs across its telephone lines.

1879   The first switchboards and operators are installed. Prior to the operators, telephone calling was *dedicated* service; telephone lines connected only two phones, and the caller could only talk with the person on the other telephone. Now the operator can connect the caller to multiple lines.

1880   Based on the work of John Carty and Thomas Doolittle, pairs of wires replace the previous single-wire telephone lines, and copper is used instead of iron. This results in cleaner sound quality that makes long-distance calling possible because the signal experiences less distortion across the distance. These two innovations are quickly adopted worldwide.

1880   All the Bell operating companies merge to form the American Bell Telephone Company.

1881   Western Union sells Western Electric to the Bell Company. Western Electric then takes over all of Bell's manufacturing.

1884   The first major regulation of telecom by a city government: A New York law requires all telephone lines to be buried underground to eliminate the unsightly cobweb of overhead telephone lines.

1885   AT&T is formed to consolidate all of the Bell Company's long-distance toll-call operations. General Manager Theodore Vail has a vision to grow AT&T into the dominant provider of all long-distance calling.

1891   Almon Strowger, a Kansas City undertaker who felt operators were steering his customers to other undertakers, invents the first dial telephone and mechanical switching system, to avoid human switching. The first installation of a nonattended switch is in 1892. New York City, however, will not be fully automated for another half-century. The installation of automated central office switches is completed in New York City in 1940.

1893   Home telephone service in New York City costs $180 a year. In many markets today, the annual cost of a telephone is little more than $180.

1897   The Independent Telephone Association (ITA) is formed. Most ITA members are small telephone companies that offer service in rural areas where Bell does not operate. Today, there are close to 1,500 independent companies, the largest of which is GTE.

1899   AT&T emerges as the parent company of all Bell operations.

1900   The Marconi International Marine Communications Company is formed. Italian Guglielmo Marconi is credited as the inventor of wireless telecommunications. The U.S. Navy immediately begins using Marconi's wireless telecom system so that its ships can communicate to shore.

1906   The vacuum tube is invented by Lee De Forest. He sells the patent rights to AT&T, which uses the technology to amplify analog voice signals that improve the call quality on long-distance calls.

1907   All research and development forces of the Bell System are consolidated to become the engineering department of Western Electric, later known as Bell Labs.

1907   New York and Wisconsin form the first state regulatory agencies.

1909   Using its superior financial position, AT&T acquires control of financially troubled Western Union. Acquiring Western Union's network significantly increases AT&T's network capacity because the same wires can be used for both telephone calls and telegraph messages. This giant telecommunications monopoly now controls all telephone and telegraph service except for the small independent telephone companies in rural areas.

1914   President Woodrow Wilson and the Justice Department require a divorce of AT&T and Western Union. Northwestern Bell, an AT&T company, is accused of not connecting with independent telephone companies. N. C. Kingsbury, vice president of AT&T, agrees to sell control of Western Union and allow all other telephone companies to use Bell System lines for long-distance calling.

1914   During World War I, numerous undersea cables are cut by German and British navies.

1915   The first coast-to-coast telephone call is made between New York and San Francisco. De Forest's vacuum tube technology is used on the call, and the copper wire is as thick as a pencil. The second transcontinental line, following a southerly route through Texas, is completed in 1923.

1918   AT&T begins using multiplexing technology to send two or more conversations across one pair of wires.

1921   The Graham Act exempts telephone companies from the restrictions of the Sherman Antitrust Act. In an effort to encourage them to expand their networks rather than focus on competition, carriers such as AT&T are allowed to operate as monopolies.

1927   AT&T first offers regular radio-telephone service between New York and London.

1930s   AT&T and Western Union begin widely offering leased-line service.

1934   The Federal Communications Commission (FCC) is established and given the charge to develop an American-controlled world telecommunications network.

1945   Western Union begins installing the first nationwide microwave radio beam communications system to replace traditional copper telecommunications lines.

1950   The first beeper is used by a New York doctor, whose golf game is interrupted by the first beeps of his new Motorola pager.

1956   The *Hush-A-Phone* settlement establishes the precedent that non-Bell devices can be attached to Bell phone lines. The Hush-A-Phone is a bowl-shaped device that callers put to their mouth to prevent those nearby from eavesdropping.

1956   In the Consent Decree, the Justice Department allows AT&T to continue as a monopoly. In exchange, AT&T has to allow other companies to access any newly patented technology developed by AT&T's Bell Labs. This makes microwave technology available, and 13 years later, Microwave Communications, Inc. (MCI) uses this to become AT&T's biggest competitor.

1962   AT&T launches Telstar 1 to offer the first satellite telecommunications link between the United States and Europe.

1962   Cellular telephone service is first tested.

1965   Centrex is first installed in Newark, New Jersey, at the Prudential Life Insurance Company.

1968   The Carterfone Decision establishes that nontelephone company devices can be connected to telephone company lines. Dallas-based Carter Electronics company develops a low-tech acoustic coupler that allows mobile radios to connect to a phone line. AT&T refuses to allow the Carterfone to be used, so the small company sues AT&T and wins.

1969   MCI begins building a microwave network to handle long-distance calls. It first offers the service in 1972. The FCC rules that incumbent telephone companies, especially AT&T, must allow MCI to connect to their network. The combination of networks begins to be called a *public-switched network.*

1969   In order for scientists and universities to share research information, the Department of Defense starts the ARPANET. In 1984, 500 computers use the system, and the term "Internet" begins to surface.

1974   The Justice Department begins an antitrust suit against AT&T for monopolistic, anticompetitive practices. A settlement is reached in 1982 and the breakup of AT&T is implemented in 1984, known simply as *divestiture.*

1975   The FCC offers the first licenses for cellular mobile telephone service.

1976   Southern Pacific Transportation Company decides to develop a telecommunications network alongside its existing railroad tracks. The new venture, called SP Communications, is later known as Sprint.

1980s   AT&T begins offering T-1 service to large organizations, such as universities and Fortune 100 companies. T-1 service allows one ordinary telephone circuit to carry 24 data or voice conversations simultaneously. The technology is developed in the 1960s as a way to increase AT&T's network capacity without building additional outside telephone lines.

1981   The FCC allows the resale of all interstate telephone service. This opens the market for hundreds of resellers, most notably Long Distance Discount Savers, or LDDS (later known as WorldCom). Within 15 years, resellers account for more than 20% of all long-distance revenue.

1984   The divestiture of AT&T is implemented. To settle the Justice
       Department's suit brought in 1972, AT&T agrees to spin off
       all seven Regional Bell Operating Companies (RBOC) [see
       Figure 2.1]. The RBOCs handle local-access services, local
       calling, and toll calling within 197 local access and transport
       areas (LATA), as well as the yellow pages. AT&T keeps interlata,
       interstate, and international long-distance calling; Western
       Electric, its manufacturing business; and Bell Labs, the research
       company.

1985   The FCC's equal access ruling requires local carriers to give their
       customers access to any long-distance carrier. AT&T's position
       as the default carrier is weakened.

1988   ISDN service is first offered.

1989   The World Wide Web is created so users can more easily access
       information on the Internet.

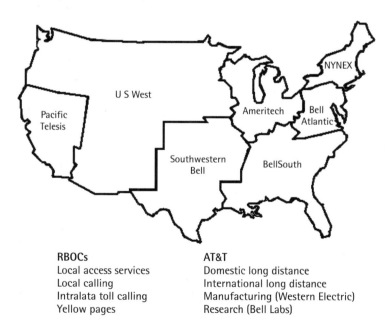

RBOCs	AT&T
Local access services	Domestic long distance
Local calling	International long distance
Intralata toll calling	Manufacturing (Western Electric)
Yellow pages	Research (Bell Labs)

Figure 2.1 The breakup of Ma Bell.

1990s   Packet-switching services such as frame relay and ATM begin to be widely offered. These services allow businesses to send large amounts of data in bursts across the carrier's network. Many businesses replace their expensive dedicated private lines with packet-switching services to cut costs.

1991   GTE acquires Contel and becomes the largest local exchange carrier in the world.

1992   AT&T acquires Western Union's Telex, teletypewriter, and e-mail services, marking the end of Western Union's position as a dominant carrier.

1994   The FCC begins auctioning operating licenses for personal communications services.

1995   AT&T divests its equipment business, Bell Labs, which is renamed Lucent Technologies. AT&T also divests its computer systems company, NCR.

1996   The Telecom Act of 1996 provides wide-ranging changes to the industry. Most of the changes are to open up all telecom markets to greater competition. Some of the effects of the act that relate to telephone service are:

- Local companies can offer long-distance service after completing a 14-point checklist.

- Long-distance carriers can offer local service.

- Local carriers must allow competitors to connect to their networks on reasonable terms.

- Universal service is established. A fee is added to each phone bill. The money goes into the Universal Service Fund. These funds are used to subsidize rural telephone companies and provide telecom services to schools, hospitals, and libraries.

- Local number portability, which allows customers to keep the same phone number when switching to a different local carrier.

1996   In order to capitalize on the changes brought about by the Telecom Act of 1996, two RBOC mergers are announced: Bell Atlantic and Nynex will merge, and Southwestern Bell and Pacific Telesis will merge.

1997   For the first time, AT&T's market share of the long-distance industry drops below 50%. AT&T's revenues continues to increase, however, because the demand for calling services is growing by about 30% per year. Voice traffic has increased due to lower calling rates brought about by increased competition. Call volume has also increased due to the prevalence of computer data traffic on phone lines.

1997   WorldCom announces an agreement to acquire MCI. This move capped off WorldCom's previous acquisition of more than 100 other telecom companies, including MFS, UUNet, CompuServe, and Brooks Fiber. The MCI merger made WorldCom the second largest long-distance provider. WorldCom's diversity gives it a great advantage in being able to provide all of a business' telecom services.

1998   Iridium first begins offering global wireless service using a network of 66 satellites. It is plagued with low sales and technical difficulties and eventually files for bankruptcy in 1999. Iridium restructures and is again offering global satellite service.

1998   Two more major mergers are announced: Bell Atlantic and GTE agree to merge, and SBC and Ameritech agree to merge.

1999   The fourth largest long-distance carrier, Qwest, announces plans to merge with RBOC U S West.

1999   Digital subscriber line service (DSL) begins to be widely offered. DSL service provides a high-speed broadband connection across ordinary copper phone lines. DSL is an affordable way for residential users to have high-speed dedicated Internet access.

This timeline provides an understanding of the big picture of telecommunications, but for more detail, numerous books are available. The definitive book on this subject is *The Story of Telecommunications* by George P. Oslin, who was a Western Union employee for more than 35 years. He experienced much telecom history first-hand, and his book is a comprehensive, well-written work.

Reference

[1]   Oslin, G. P., *The Story of Telecommunications*, Macon, GA: Mercer University Press, 1992.

3

The complete do-it-yourself telecom audit

Anyone can successfully perform a telecom audit. If you are already familiar with your own services and willing to spend a couple hours a month reviewing your bills, you can trim your telecom costs by hundreds or thousands of dollars each month. In this chapter, I will explain a simple step-by-step method for performing a complete audit of all your telecom services. Many professional telecom consultants and bill auditors use this same approach.

Numerous phone bill auditing firms have sprung up in the last 10 years. These firms provide a valuable service for businesses that lack the time and expertise to audit their own phone bills. These consultants are either paid an hourly rate or a percentage of the savings and refunds that they generate. They prefer to work with customers whose monthly phone bills total $5,000 to $50,000. Smaller companies keep a close eye on their expenses, and larger corporations have a full-time telecom staff that manages the services each month. When customers keep their phone bills clean, there is less opportunity for the consultant. But no matter what size your company, you need a strategy for controlling and minimizing your telecom expenses.

The audit method explained in this chapter is a comprehensive system to manage all of your telecom expenses. This system is scalable. You may not have the time to perform a complete audit; you may only be concerned about one of your services. In that case, you can use this book as a reference tool and selectively attack one telecom service at a time. But if you have time and follow this plan, you will know exactly what services your organization uses, as well as the exact cost for those services. You will then be able to cut your costs by thousands of dollars.

The audit objective

Anyone can successfully perform a telecom audit on his or her own phone bills if given the proper game plan. But you must first know your objective. For a complete audit, the objective should be to:

- Create a complete *inventory* of current telecom services.

- Show the *current cost* of those services.

- *Make changes* to the telecom environment so it will become more efficient and cost effective.

You should also decide how to manage the information and how to report your audit's results. To whom will you report? Will you use simple spreadsheets, or will you give a formal presentation using graphs, charts, and slides?

Preparation

Before beginning the audit, start out slowly and stay organized. Soon you'll be swimming in phone bills and you may get lost in all the details. Before doing anything else, gather all the phone bills and related records in one central location. If you have multiple locations, you must inform each location to send in all their records, and you must budget time for this. The following records should be gathered for each location:

- Local phone bills;

- Customer service records from local exchange carriers (LECs) (see Chapter 10);

- Long-distance phone bills;

- Wireless phone bills;

- Paging bills;

- Data bills;

- All proposals;

- Term contracts;

- Internal list of phone numbers;

- Wide-area network (WAN) diagram.

Inventory your records

In a spreadsheet, create an inventory list of all the records. Start your list by requesting all of the account numbers from your accounts payable department. The inventory will help you keep track of what records are still outstanding, and this spreadsheet will allow you to sort and group the data by location or by service. Table 3.1 shows a list of all the accounts for Acme Manufacturing, a fictional company. After presenting my clients with their

Table 3.1   List of Telephone Accounts for Acme Manufacturing

Location	Vendor	Service	Account Number	Bill Date	Bill Amount
St. Louis	SBC Communications	Local	314-555-XXXX-001	11/15/01	$5,505.20
St. Louis	SBC Communications	Local	314-555-XXXX-001	11/15/01	$1,250.10
St. Louis	AT&T	Long distance	171-XXX-XXXX-001	11/10/01	$4,675.35
St. Louis	AT&T	Long distance	151-XXX-XXXX-001	11/10/01	$603.64
St. Louis	AT&T	Data	1001-XXX-XXXX-001	11/20/01	$1,510.00
St. Louis	AT&T Wireless	Cellular phones	314-555-XXXX	11/3/01	$1,255.00
St. Louis	PageNet	Pagers	XXXXXXX	11/5/01	$350.65
			Total monthly expense:		$15,149.94

own account list, they usually tell me this is the first time they have ever seen the actual cost of their monthly telecom expenses, and they are surprised at the high expense.

Audit phone bills

Once you have gathered all the records, you will have a mountain of paperwork in front of you. Now you are ready to audit. It works well to keep the phone bills organized by location or according to service: local, long distance, data, or wireless. Before doing a detailed line-by-line audit (which most people never have time for) do a cursory review of the bills. Take a common sense approach and look at the bills one by one. Try to get an idea of where your organization spends most of your telecom dollars. Then you will be ready to start auditing each bill.

First read the summary pages of each bill and see exactly what the carrier is charging you for. Does your organization really need all of these services? Do the charges seem correct? Are the taxes and fees excessive? Next, spot-check the call detail to get a ballpark idea of the cost per minute. When you encounter a charge that you do not understand, call the phone company and question it about the charge. Most phone company employees will patiently explain the phone bill to you, and many will offer suggestions on how to reduce the bill. You can also get end users involved. Your employees who are actually using the phones can be a valuable source of information. Ask them if they have any ideas on how to reduce the cost of your telecom services.

Monthly savings

The desired result of a telecom audit is to *reduce monthly expenses* by doing one or more of the following:

- *Correct erroneous rates,* such as wrong per-minute rates on a long-distance bill.

- *Close unused accounts,* such as local phone lines that are no longer used.

- *Eliminate employee expenses,* such as a mobile phone expenses for ex-employees.

- *Eliminate unnecessary services,* such as pager replacement programs and wire maintenance plans on local lines.

- *Add missing discounts* back to accounts and secure refunds of the overcharges, such as a missing association discount on a long-distance bill.

- *Eliminate excessive fees* by having the carrier waive them, such as monthly fees for 800 numbers on a long-distance bill.

- *Add missing promotions* back to accounts, such as missing waived local loop charges on a dedicated private line.

- *Negotiate new promotions,* such as free nights and weekends on a mobile phone.

- *Implement lower pricing on an existing service,* such as by changing rate plans on a mobile phone.

- *Change to a more cost-effective service,* such as replacing cellular phones with personal communication service (PCS) phones.

- *Negotiate a new contract* with your current carrier.

- *Change vendors,* for example, move intraLATA calls from the local carrier to the long-distance carrier.

One-time credits and refunds

A successful telecom audit will not only reduce ongoing monthly expenses but also produce one-time refunds and credits such as the following:

- *Refunds of overcharges.* A customer is always entitled to a full refund if the phone company has erroneously billed the customer. The refund should always include the taxes paid, and sometimes the carrier will also refund interest.

- *Bonuses.* Many carriers will give bonus credits to customers who sign term agreements. One of the most common bonuses is a credit in the 13th month of a 24-month term agreement with a long-distance carrier.

- *Installation credits.* Many carriers will reimburse any expense a customer pays to have the carrier's service installed. For example, a customer who signs up for new frame relay service will have to buy routers. Once the frame relay service is up and running, the carrier will put a credit on the customer's account to reimburse that customer for the router purchase.

◆ *Promotions.* From time to time, such as once a quarter, carriers give away promotional credits. A typical promotion will be something like *waived local loop charges* for 6 months. During the course of a telecom audit, the auditor should ask all the carriers (especially the long-distance carriers) to give the customer the next promotion. The carriers are most likely to give new promotions when the customer is negotiating a new term contract.

◆ *Courtesy credits.* Most customer service representatives are able to credit customer accounts to satisfy minor discrepancies. For example, if a carrier cannot speedily implement a new rate plan on a mobile phone, it may give a $50 one-time courtesy credit as a good faith gesture. It does not hurt to ask for courtesy credits, especially if the carrier owes you a favor.

Tables 3.2 through 3.8 show the most common areas where you may be able to reduce expenses. Cross-reference each phone bill with the cost management tips listed. If you are unfamiliar with the cost management tip, turn to the chapters mentioned in the tables and familiarize yourself with these money-saving strategies.

Local service

Audit the local phone bills first. Local bills show how many lines each location has, which quickly gives you an idea of the overall telecom picture for each location. Local bills also contain a wide variety of charges that are quickly and easily reduced, such as fees and unnecessary services.

In addition, local bills have charges for local calls, intralata calls, and long-distance calls. These expenses are easily reduced, and you can have the satisfaction of having an immediate impact with your audit. Long-distance calling on local bills, commonly called "loose traffic," occurs on about one-third of all local bills. Loose traffic is very common, very expensive, and very simple to correct. Table 3.2 contains the most common cost management areas to look for when auditing local bills.

Long distance

After local bills, audit long-distance bills next. Long distance is one of the largest monthly expenses for a business, and long-distance pricing is unquestionably the most negotiable of all telecom services. Auditing the bill may turn up a few low-dollar savings opportunities, such as

Table 3.2   Local Service Checklist

	Cost Management Strategy	Page Number In This Book
1	Upgrade to Centrex	47
2	Replace WATS lines	50
3	Use a T-1 for local service	50
4	Lower line charges by moving to measured rate service	52
5	Use a CLEC	53
6	Eliminate unused lines	53
7	Reduce the number of trunks	54
8	Eliminate unused optional services such as voice mail	54
9	Replace remote call forwarding with 800 service	57
10	Cancel wire maintenance	57
11	Use an alternate vendor for voice mail	58
12	Block pay-per-use features	59
13	Reduce the cost of directory services	60
14	Lower local calling charges by changing to flat rate service	62
15	Use off-peak calling	63
16	Enroll in a local calling package	64
17	Use local call volume for greater discounts	65
18	Switch local calls to a new vendor	65
19	Use an off-premise extension	67
20	Use auto-dialers to switch intralata calls to a new carrier	70
21	Dial-around to switch intralata calls to a new carrier	71
22	Program your PBX to switch intralata calls to a new carrier	71
23	L-PIC your intralata traffic to a new carrier	71
24	Use an LEC calling plan to lower intralata call rates	72
25	Enroll in an extended local calling area plan	74
26	Move loose traffic off local bills to a long-distance bill	81
27	Loose traffic rerate credits	82
28	Full recourse	84
29	Block collect calls	85

Table 3.2   (continued)

	Cost Management Strategy	Page Number In This Book
30	Block 900 and 976 calls	85
31	Eliminate excessive fees (cramming)	86
32	Eliminate payphone bills	87
33	Eliminate bills for semipublic payphones	88
34	Receive payphone commissions from an operator service provider (OSP)	88
35	Use a customer-owned payphone	89
36	Reduce the risk of loose traffic by correcting PIC codes on CSRs	97
37	Reduce the risk of loose traffic by correcting L-PIC codes on CSRs	98
38	Eliminate extra end user common line charges	98
39	Correct erroneous tax rates	98
40	Change incorrect hunting sequence	99
41	Cancel hidden wire maintenance charges	99
42	Correct erroneous data circuit mileage	99
43	Cancel hanging data circuits	100
44	Correct term plan errors	101
45	Correct sliding scale line rates	101
46	Eliminate loose calling cards	101
47	Cancel unused voice lines and data circuits	102

eliminating fees on an 800 number. But the high-dollar opportunities are in implementing greater discounts and lower rates.

Check the current rates and discounts to make sure they match the carrier's original proposal. If there is a discrepancy, immediately notify your carrier to have the error corrected and to negotiate a refund. Whether or not the current pricing is correct, you should still try to negotiate a new contract with lower pricing. This is the highlight of a telecom audit, because significant savings can be implemented. Many consultants will not accept a new client if the client will not allow the consultant to renegotiate the client's long-distance contract.

Table 3.3 contains the most common cost management areas for long-distance service.

Table 3.3   Long-Distance Service Checklist

	Cost Management Strategy	Page Number In This Book
1	Use a discount calling card vendor	118
2	Use prepaid calling cards	119
3	Use rechargeable prepaid calling cards	119
4	Use 800 numbers instead of calling cards	119
5	Change carriers for inbound long distance	121
6	Eliminate fees for toll-free numbers	122
7	Verify PICC rates	129
8	Implement tax exemptions, if applicable	130
9	Use off-peak calling	138
10	Increase discounts with volume agreements	142
11	Request automatic monthly discount upgrades	143
12	Avoid shortfall penalties	145
13	Increase discounts with term agreements	146
14	Request association discounts	147
15	Request international discounts	148
16	Request referral programs	148
17	Request points programs	148
18	Add missing discounts	148
19	Add missing discounts back to subaccounts	149
20	Set up a national account	149
21	Move from a national account	151
22	Change from switched long distance to dedicated	155
23	Remove unnecessary voice T-1s	157
24	Move loose switched lines to dedicated	157
25	Correct virtual private network rate errors	158
26	Use tie lines to eliminate long distance calls	158

Data

Because of the technical nature of data networking services, many telephone bill auditors are intimidated by these services. Data networking technologies are complex. Modifying the existing technology or switching to another technology may reduce the cost, but most telecom consultants do not possess the technical expertise for such changes. Most consultants specialize in auditing and negotiating, not telecom technology.

However, if cost management is a customer's concern, there are many opportunities to cut monthly bills. Table 3.4 contains the most common cost management areas for data service. In some cases, the auditor may want to consult with a technician, but first try to implement some of the cost management measures documented in Chapters 15 and 16.

Wireless phones

Wireless telephone service is relatively simple. The technology is not complex, and the billing and pricing are fairly straightforward. Over the past few years, however, many more choices have been made available to customers, which means they have numerous options available to reduce their

Table 3.4 Data Service Checklist

	COST MANAGEMENT STRATEGY	PAGE NUMBER IN THIS BOOK
1	Eliminate outdated e-mail systems	177
2	Cancel bogus Internet billing	178
3	Replace missing discounts	183
4	Correct mileage errors	184
5	Disconnect dead circuits	185
6	Consolidate circuits	186
7	Volume discounts	187
8	Term discounts	187
9	Updated pricing	188
10	Eliminate ISDN loose traffic	191
11	Replace dedicated lines with frame relay	196
12	Move voice calls over frame relay	198
13	Move voice calls over the Internet	199

wireless telephone bills. Table 3.5 will help you look for the most common cost management areas for wireless service.

Paging

Pagers are relatively low-tech and inexpensive devices, so many businesses ignore this expense. But to perform a comprehensive audit, these expenses must also be reviewed. Because of the simplicity of the service, reducing

Table 3.5   Wireless Phone Service Checklist

	COST MANAGEMENT STRATEGY	PAGE NUMBER IN THIS BOOK
1	Free wireless phones	217
2	Change low-end users to emergency rate plans	220
3	Downgrade rate plans for low-end users	220
4	Upgrade rate plans for high-end users	220
5	Use a carrier that automatically upgrades your rate plan	221
6	Make last-minute rate plan changes	222
7	Secure customer promotions	225
8	Upgrade from analog to digital	226
9	Start a small group rate plan	227
10	Start a corporate account	228
11	Use someone else's corporate account	229
12	Eliminate employee fraud and waste on corporate accounts	229
13	Move high-end users off the corporate account	229
14	Use prepaid wireless	230
15	Implement an association discount	230
16	Eliminate unnecessary features	231
17	Eliminate roaming charges with a one-rate plan	236
18	Use the dual-NAM feature to eliminate roaming	237
19	Negotiate a lower rate on wireless long distance	238
20	Switch wireless long distance to the long-distance carrier	238
21	Use calling party pays to eliminate the cost of incoming calls	239
22	Use a carrier that gives free first minutes	241
23	Add free nights and weekends	241

paging costs is easy. Table 3.6 contains the most common cost management areas for paging.

Contracts

For greater detail on contract negotiation, see Chapter 20. In general, however, the following steps should be taken:

- Evaluate current contract terms.
- Audit the bill to see if the contracted pricing is given.
- Correct the pricing if needed, and negotiate a refund.
- Negotiate lower pricing and more favorable contract terms.

To learn the current market's pricing, hire a consultant, compare your contract to other businesses, or simply get competitive bids from other carriers.

Before renegotiating a contract, it is a good idea to implement all of the small changes that your audit has warranted. Correct all of the pricing errors, implement new pricing plans, and cancel unneeded services before working on a new contract. If you give your carrier too much work at once, the representative may feel overwhelmed and make mistakes because of the pressure. Toward the end of your audit, it is time to review the contracts.

Implement your changes

Because of their experience, professional phone bill auditors can easily find savings opportunities for their clients. They have done the work before; they know what to look for, so they can quickly recommend cost-cutting changes. But it is not easy for consultants to implement their

Table 3.6   Paging Service Checklist

	Cost Management Strategy	Page Number In This Book
1	Replace tone pagers with digital pagers	244
2	Purchase pagers to eliminate monthly rental charges	249
3	Cancel pager replacement programs	250
4	Use quarterly or annual billing to lower rates	251
5	Secure lower pricing with a term agreement	252
6	Consolidate all pagers to one carrier for volume pricing	252

own recommendations. Before making any changes, telephone companies always require the consultant to present a letter of authorization (LOA) signed by the client. The LOA functions like a power of attorney.

Consultants also have a reputation of bullying telephone company employees, and because fraud is so prevalent in the industry, phone company employees are very reluctant to work with consultants. Telephone company customer service representatives generally have a large number of orders on which they are working, and they are not happy to receive more work. When they are overworked, they make mistakes. Sometimes when you request a change to your account that will cut costs, phone company representatives instead make a change that increases your cost.

Auditing a $10,000-per-month customer's bills may take 3 hours, but implementing the changes may take 3 to 4 months. Most consultants are paid 50% of the savings once they are realized on phone bills. But if a client is willing to implement the consultant's recommendations on its own, many consultants would gladly reduce their fees. They may even cut them in half. As a consultant, I would much rather make $5,000 in 3 hours than $10,000 over the course of 6 months. If you have hired a consultant, you can cut his or her fees if you are willing to handle all of the implementation.

Once all the orders are placed with the phone companies, be sure to verify all implementation. After placing the order, call the carrier back to double-check that the changes have been made. Diligently check the next phone bill to ensure that the new pricing is in place.

Document the audit

Keep copies of all documents related to the audit, including the phone bills, carrier correspondence, proposals, and contracts. After doing the legwork of completing a telecom audit, you will also want to be able to clearly report the results of your work. Professional phone bill auditors usually document each recommended change in a simple one-page spreadsheet that follows this format:

$$\text{old cost} - \text{new cost} = \text{monthly savings}$$

Many of the examples in this book follow this format.

Table 3.7 shows an example of a typical report used by a telecom consultant to present recommendations to a client. The report first explains the change in simple language, then shows the cost comparison. In this particular example, the business can save money by canceling a stand-alone AT&T long-distance account and moving the traffic to the main AT&T long-distance account. Table 3.8 is a simple report of a one-time refund. In this fictional example, AT&T has given Acme a refund of the overcharges associated with having a stand-alone account.

Table 3.7   Monthly Savings Moving Long-Distance Traffic to Main Account

SITUATION:					
Long-distance calls from two new fax lines are billing on a separate AT&T account.					
RECOMMENDATION:					
Close the account and move the lines to the main AT&T account.					
	CURRENT MINUTES	CURRENT CPM	CURRENT COST	NEW CPM	NEW COST
Intralata	1,502.0	$0.125	$187.75	$0.080	$120.16
Intrastate	345.4	$0.125	$43.18	$0.090	$31.09
Interstate	2,351.4	$0.150	$352.71	$0.100	$235.14
Account fees:			$20.00		$0.00
Total:			$603.64		$386.39
Monthly savings:					$217.25
Annual savings:					$2,606.99

CPM = cost per minute

Table 3.8   One-Time Credit Refund of Long Distance

SITUATION:	
Long-distance calls from two new fax lines were billing on a separate AT&T account.	
The account has been closed and a 3-month refund of the overcharges has been negotiated.	
The credit will appear on the December invoice.	
Total refund:	$1,810.91

4

Local service

In the early years of the telecom industry, customers received their telecom services from one single provider; usually AT&T in highly populated areas, or one of the independent telephone companies in rural areas. After divestiture, the monopoly of AT&T was broken up and end users had two carriers to deal with: the LEC for local calls and the interexchange carrier (IXC) for long-distance calls. Customers then had two different companies to deal with, and they received two separate phone bills. This chapter outlines the services typically provided by the local exchange carriers and includes a sample local telephone bill.

Local access and transport areas

What determines if the telecom service is local or long distance? As part of the agreement between AT&T and the Justice Department at the time of AT&T's divestiture, this question was one of the most important to be answered. Why? The negotiators on the LEC side of the table and on the AT&T side of the table haggled over who got to carry which calls. Ultimately, the question was: "Who gets to earn the revenue from these calls?"

A huge amount of money was at stake. In the end, it was decided to divide the country into 777 LATAs.

LATA boundaries were determined by population and, in most cases, LATAs are the size of two to five counties. In some cases, the LATA is almost the same geographically as the area code, but they are not the same. Most LATAs are contained within a state, but some cross state lines. The Chicago LATA, for example, includes Gary, Indiana. A call from Gary to Chicago is an intralata call, even though it is an interstate call.

In 1984, as part of the agreement between the Justice Department and AT&T, local telephone companies would provide all telephone lines (*access*) and carry all telephone calls (*transport*) within the LATA. Calls between LATAs, or interexchange calls, were to be carried by long-distance carriers. Figure 4.1 shows the two LATAs in Washington state. A call from Spokane to Seattle to Tacoma is an intralata call and should be carried by the local carrier. A call from Seattle to Spokane, however, is an interlata call and is to be carried by the long-distance carrier.

Although the breakup of AT&T was designed to benefit consumers because competition would bring prices down, the situation initially proved to be very confusing for consumers. It was unclear to customers which company was responsible for which service. This was very frustrating, especially for customers whose lines needed to be repaired. Bell and AT&T often blamed each other. Meanwhile, the customer lost valuable

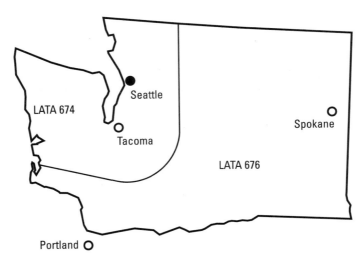

Figure 4.1   Local access and transport areas in Washington state.

phone calls while the carriers were busy pointing fingers instead of repairing lines. Today, local carriers and long-distance carriers still bicker about who is responsible when a customer experiences technical difficulties with the phone lines.

Where does the phone company's network start?

The demarcation point is usually at the wall jack on the outside of your building, or the RJ21X telephone jack inside your equipment room. It is at this point that the phone company responsibility to maintain the line ends, and the customer's responsibility begins.

The Telecom Act of 1996 and local service

The Telecom Act of 1996 allows long-distance carriers to provide local lines and handle local calling. It is very expensive for long-distance carriers to build local lines from their central offices all the way to a customer's premise, so most customers still cannot choose a new local carrier. Most competitive local exchange carriers (CLEC) only offer facilities-based local service to large corporate customers. CLECs are more likely to offer local service to small businesses under a reseller arrangement (see Chapter 5 for more information on CLECs).

The Telecom Act of 1996 also allows local carriers to offer long-distance service. This is fairly easy for local carriers. All they have to do is resell the services of a long-distance carrier. Customers can have simplified telecom billing—their long distance can be billed on their local bill. It is therefore easier for local carriers to enter the long-distance market than it is for long-distance carriers to offer local service. To level the playing field, the Telecom Act of 1996 requires local carriers to first open their market to competition, prior to entering the long-distance market themselves. To prove that they have opened their market to competition, they must fulfill a 14-point checklist.

Local telephone companies

Local telephone companies are known by a variety of names according to their function. The following acronyms and terms are most commonly used to describe telephone companies.

LEC   Local exchange carrier. Every local phone company is an LEC, for example, Cincinnati Bell Telephone Company.

IXC   Interexchange carrier. Long-distance companies carry calls between exchanges (or LATAs) and are known as IXCs. One of the largest wholesalers of long-distance service is a company whose name is IXC.

BOC   Bell Operating Company. The 22 local telephone companies that were originally owned by AT&T were transferred to seven Regional Bell Operating Companies. Mountain Bell, Northwestern Bell, and Pacific Northwest Bell were all transferred to the RBOC U S West.

RBOC   Regional Bell Operating Company. At the time of divestiture, AT&T spun off 22 BOCs, which were managed by the seven RBOCs. Since then, the RBOCs and GTE have dominated the local service marketplace. The seven RBOCs were Ameritech, Bell Atlantic, BellSouth, NYNEX, Pacific Telesis, Southwestern Bell, and U S West. Southwestern Bell (now known as SBC Communications) has merged with Pacific Telesis and Ameritech. Bell Atlantic has merged with NYNEX and GTE. Now, there are only four RBOCs.

ILEC   Independent local exchange carrier, a local carrier that is not a part of the Bell System. GTE long boasted that it is the "largest independent LEC."

CAP   Competitive access providers. These small entrepreneurial companies provide "access" to the public-switched network and bypass the incumbent LEC, such as Teleport Communications Group in the 1980s.

CLEC   Competitive local exchange carrier. After the Telecom Act of 1996, a multitude of CLECs started up, such as Adelphia, Nextlink, and SNiP. Due to changing market conditions and financial mismanagement, numerous CLECs have filed for bankruptcy, such as e.spire and Teligent. In addition to long-distance service, AT&T now offers local service in many markets, so AT&T can be called a CLEC.

ILEC   Incumbent local exchange carrier. If a CLEC such as Nextlink is operating in BellSouth's region, Nextlink will refer to BellSouth as the incumbent carrier.

The local bill

Items on a local bill fall into one of four categories: regulated charges, non-regulated charges, taxes and fees, and charges from other carriers.

Regulated charges

Tariffs are filed with the state Public Utility Commission (PUC). Tariffs define the rules and pricing of telecom services. The prices for regulated charges are nonnegotiable, although the carriers often file additional tariffs to be used to offer special pricing to large customers. Line charges and calling rates are regulated charges.

Nonregulated charges

Carriers are not required to file tariffs for these services. Prices on nontariffed items are set based on current business and competitive conditions for the area. These charges are often "nonessential" services such as calling features and voice mail.

Taxes and fees

There are typically four types of taxes and fees that appear on local phone bills:

- Service fees and charges, such as the 911 surcharge and PUC funding fees;

- Franchise tax—usually a local item like a municipal charge;

- Sales, use, or special taxes—usually written as "state and local taxes" on the bill;

- Federal excise tax—this 3% tax was originally a World War II emergency tax.

Charges from other carriers

Charges from companies other than your own local carrier sometimes appear at the back of your local telephone bill. Because local carriers keep a small portion of the money, they are always happy to provide this service to other companies. Some of the charges that may appear are collect calls, 900 calls, voice mail, long distance, and Internet access.

Sample local telephone bill

Figure 4.2 is a typical local bill for a fictional business in Florida. Like other local bills, this one follows an outline: summary page, monthly service, local calling, intralata calling, and charges from other carriers.

Summary page

The bill's first page usually has two sections: the summary of charges and the bill remittance page. The summary of charges outlines the current charges, any past-due charges, charges from companies other than the local

---

Telephone Company A

Page 1

Acme Manufacturing
Account Number: 561-555-5555-001
Bill Period Date: Jun 27, 2001

Thank you for choosing Telephone Company A. We sincerely appreciate your business.

Summary of Charges

Previous Charges	*Amount*
Amount of Last Bill..	325.12
Less Payments...	325.12
Balance - Thank you for your payment...	.00

Current Charges	*Amount*
Telephone Company A	
Monthly Service Charges...	185.07
Local Usage..	12.54
Other Charges and Credits...	2.20
Itemized Calls..	43.44
Optional Calling Services..	-6.52
Taxes..	40.94
Total Current Charges for Telephone Company A........................	277.67
Charges for Other Companies...	31.04
Total Current Charges Due Before Jul 23.............................	308.71
Total Amount Due...	308.71

Important Notice(s)

Late Charge Reminder: A 1.5% Late Payment Charge will apply to any unpaid balance as of Jul 23.

Nonpayment of Regulated Charges may result in discontinuance of service. Failure to pay unregulated and certain other charges will not

result in an interruption of local service. The amount of Regulated Charges may be obtained by calling 555-555-5555.

- -

Please make check payable to Telephone Company A in US funds.

Jul 23, 2001	308.71	Account Number:	561-555-5555-001
		Bill Period Date:	Jun 27, 2001

Mail to:
Telephone Company A   Acme Manufacturing
123 Main Street   123 Main Street
Boca Raton, FL XXXXX   Boca Raton, FL XXXXX

Figure 4.2 Sample local telephone bill.

Telephone Company A

Acme Manufacturing
Account Number: 561-555-5555-001
Bill Period Date: Jun 27, 2001

Helpful Numbers
Telephone Company A
NOTE: Numbers for other companies are listed on their pages.

Billing Questions or to Place an Order:	800-555-5555
Repair:	305-555-5555
Text Telephone (TTY) Users:	800-555-5555

Detailed Statement of Charges

Monthly Service Charges

Monthly Service - Jun 27 thru Jul 26

Basic Services	Quantity	Unit Charge	*Amount*
1. Business Line	2	43.87	87.74
2. Emergency 911 Charge. This charge is billed on behalf of Dade County.			
3. FCC Charge for Interstate Access	1	#	.45
4. Telecommunications Relay Service	2	8.14	16.28
	2	.10	.20
Total Basic Services..........			104.67

Optional Services	Quantity	Unit Charge	
5. Three-Way Calling	1	4.95	4.95
6. Call Forwarding	1	4.95	4.95
7. Directory Advertising	1	see detail	70.50
Directory: FT LAUDERDALE, FL	70.50		
Total Optional Services.........			80.40
Total Monthly Service Charges.........			185.07

Local Usage

Local Usage Summary for 561-555-5555
Local Measured Usage to the Expanded Area

	Calls	Total Mins	Charges	*Amount*
	55	153	12.54	
8. Local Usage Summary				12.54
Total Local Usage..........				12.54

(continued) ▶

Figure 4.2 (continued)

carrier, and the total amount due. The perforated bill remittance page is to be included with the check when the customer sends in the payment.

When looking at this page of the bill, a seasoned bill auditor notices that the customer has measured local service. It is billed $12.54 for local calls. It may be cheaper to switch to a flat-rate service. The customer also has intralata calls. (On Telephone Company A's bills, intralata calling is listed as "Itemized Calls.") Any usage, whether it is local, intralata, or long distance, that appears on a local bill can almost always be reduced. In this case, the customer also has $31.04 listed under "Charges for Other Companies." These charges can almost always be reduced or removed altogether.

Page 3

Telephone Company A

Acme Manufacturing
Account Number: 561-555-5555-001
Bill Period Date: Jun 27, 2001

Detailed Statement of Charges

Other Charges and Credits	**_Amount_**
Directory Assistance (DA) Usage	
9. 4 Call(s) to Local DA at No Charge..	.00
10. 7 Call(s) to Local DA at $0.25 Each..	1.75
11. 7 Call(s) to Local DA at $0.25 Each..	.45
Total Other Charges and Credits...	2.20

Itemized Calls   **_Amount_**

561-555-5555
Direct Dialed Calls

	Date	Place Called	Number Called	Rate	Time	Min	
12.	05/29	WPALMBEACH FL	561-XXX-XXXX	HD &	09:24 AM	5.0	1.20
13.	05/29	WPALMBEACH FL	561-XXX-XXXX	HD &	03:51 PM	12.0	2.88
14.	05/30	WPALMBEACH FL	561-XXX-XXXX	HD &	05:45 AM	18.0	4.32
15.	06/02	BIG PINE FL	561-XXX-XXXX	HD &	12:51 PM	34.0	8.16
16.	06/03	WPALMBEACH FL	561-XXX-XXXX	HD &	03:21 PM	21.0	5.04
17.	06/05	WPALMBEACH FL	561-XXX-XXXX	HD &	02:51 PM	3.0	.72
18.	06/10	SUGARLFKEY FL	561-XXX-XXXX	HD &	12:57 PM	6.0	1.44
19.	06/12	WPALMBEACH FL	561-XXX-XXXX	HD &	12:57 PM	29.0	6.96
20.	06/20	BIG PINE FL	561-XXX-XXXX	HD &	11:31 AM	53.0	12.72

Total Direct Dialed Calls.. 43.44
Total Charges for 561-555-5555... 43.44

The above total does not include the following taxes:

Federal Tax..	$1.30
State/Local Tax.......................................	$6.52
Florida Gross Receipts Surcharge..........................	$0.43

Total Itemized Calls.. 43.44

& Charges Included in Discount Summary.

Optional Calling Services   **_Amount_**

Discount Summary

21. $43.44 of Calls at 15% Discount	-6.52
Total Business Saver Service...	-6.52
Total Optional Calling Services...	-6.52

(continued) ▶

Figure 4.2 (continued)

Monthly service

Page 2 of the bill shows all of the monthly recurring charges billed by the local carrier. The first item is usually the charge for the lines, which is a fixed expense. Other charges appearing in this section are any optional services, such as call forwarding, wire maintenance, hunting/rollover, and directory advertising in the yellow or white pages. In the sample bill, the customer has two lines that cost $43.87 each. This is a high rate for measured service, so a bill auditor would research why the line rate is so high. A bill auditor would also check with the end user to make sure the three-way

```
                                                                      Page 4
Telephone Company A                  Acme Manufacturing
                                     Account Number:      561-555-5555-001
                                     Bill Period Date:    Jun 27, 2001
                        Detailed Statement of Charges

Taxes                                                               Amount
Taxes on Regulated Services
    22. Federal Tax...................................................     7.10
    23. State/Local Tax..............................................    17.64
    24. Florida Gross Receipts Surcharge............................     2.36
    25. County Tax..................................................    13.84
Total Taxes on Regulated Services.....................................   40.94
Total Taxes...........................................................   40.94

Total Telephone Company A Current Charges.............................  277.67


                                                            (continued) ▶
```

Figure 4.2 (continued)

calling and call forwarding features are actively used. If not, they should be canceled.

Local calling

As previously stated, local lines can be billed one of three ways: flat-rate service, measured-rate service, or message-rate service. With flat-rate service, local calls are not charged, so this section of the bill may be missing. In some cases, the local carrier will still provide a summary of local calls even though it is not charging for the calls. With measured-rate or message-rate

```
                                                                          Page 5
Telephone Company B                          Acme Manufacturing
                                             Account Number:        561-555-5555-001
                                             Bill Period Date:      Jun 27, 2001

        For Telephone Company B Billing Questions, Call 1-800-555-5555
                           Detailed Statement of Charges

                                                                       Amount

                       Number Called   Rate    Time     Min
                       515-XXX-XXXX     HD     03:36 PM   15              3.75
                       404-XXX-XXXX     HD     12:51 PM   4               1.00
    28.  05/30  SALEM OR        503-XXX-XXXX   HD  03:21 PM   9           2.25
    29.  06/02  TAMPA FL        813-XXX-XXXX   HD  03:36 PM   24          6.00
    30.  06/03  DESMOINES IA    515-XXX-XXXX   HD  03:51 PM   12          3.00
    31.  06/05  DESMOINES IA    515-XXX-XXXX   HD  05:45 AM   3            .75
    32.  06/10  ATLANTA GA      404-XXX-XXXX   HD  05:45 AM   6           1.50
    33.  06/12  ATLANTA GA      404-XXX-XXXX   HD  03:21 PM   29          7.25
    34.  06/20  TOLED OH        419-XXX-XXXX   HD  06:00 AM   1            .25
Total Direct Dialed Calls.................................................         25.75
Total Charges for 561-555-5555...........................................         25.75
The above total does not include the following taxes:
         Federal Tax........................................     $0.77
         State/Local Tax....................................     $3.86
         Florida Gross Receipts Surcharge...................     $0.66

Total Itemized Calls.....................................................          25.75

Taxes                                                                  Amount
Taxes on Regulated Services
    35. Federal Tax.........................................................        .77
    36. State/Local Tax.....................................................       1.92
    37. Florida Gross Receipts Surcharge....................................        .66
    38. County Tax..........................................................       1.95
Total Taxes on Regulated Services..........................................       5.29
Total Taxes................................................................       5.29

Total Telephone Company B Current Charges..................................       31.04
```

This portion of your bill is provided as a service to Telephone Company B. There is no connection between Telephone Company A and Telephone Company B.

Figure 4.2 (continued)

service, this section of the bill itemizes the local calling usage and the charges associated with that usage. If the customer switches to flat-rate local service, the $12.54 charge for local calls, listed on page 2 of the bill, will be eliminated.

Intralata calling

In this bill, intralata calls are listed under the heading "Itemized Calls" on page 3 of the bill. The customer is paying $0.24 per minute. The business saver service discount plan gives the customer a 15% discount. This lowers the effective rate to $0.20 per minute, which is still far too high for today's marketplace. The customer should consider switching these calls to its long-distance carrier. Most long-distance carriers will carry intralata traffic for less than $0.10 per minute. Telephone Company A can implement this change in its central office. Before switching, the customer should first try to negotiate a lower rate with Telephone Company A.

The next heading on page 3, "Optional Calling Services," simply shows the business saver service discount plan. The customer receives a 15% discount on intralata calling. If the customer had 800 service or calling cards through Telephone Company A and was using these services within the LATA, the 15% discount would also apply to these calls. However, most customers use their long-distance carrier for calling cards and 800 service.

Charges for other companies

After the local carrier has itemized its charges for monthly service, local calling, and LATA calling, the remainder of the bill is comprised of charges from other companies. The most common charges in this section, also listed on page 3 of the bill, are for legitimate services such as Internet access, interlata calling card charges, direct dial long distance, and collect calls. Fraudulent charges billed by other carriers such as United States Billing, Inc. (USBI) and Hold, Inc. usually appear in this section of the bill. Slamming, cramming, and 900 calls, if billed, will appear in this last section of the local bill.

In the sample bill, one of the two lines is showing long-distance calls carried by Telephone Company B at $0.25 per minute. The true cost-per-minute is actually higher, because the calls are billed in full-minute increments. Full-minute billing is about 8% more costly than billing the same calls in 6-second increments (see Table 7.1 for a further explanation).

This customer might have been slammed by Telephone Company B, but it is more likely that a customer error caused this problem. Assuming the customer has a separate Telephone Company B long-distance account, this line should have been billed on that account, not on the local bill. But if the customer failed to inform its Telephone Company B account team about this line when it was first ordered, the line will bill alone on the local bill. This situation is called "loose traffic" because the calls (traffic) on this line are billing apart (loose) from the main Telephone Company B account. To correct this problem, the customer should inform Telephone

Company B and request a refund. Telephone Company B will request bill copies and then "rerate" the bill—it will recalculate the bill at the correct lower rates and refund the difference.

Now that we have examined the outline of the local bill, Chapter 5 explains, in detail, the different local line charges that appear on the local bill. Customers have many choices when it comes to their local service.

Only through systematic review and diligent auditing of your local bills can you avoid being overcharged. If a customer blindly accepts the carrier's service offerings and billing, without question, the company will be subject to overcharges and inefficiencies every month. The culture of the telecommunications industry is built on carriers providing customers whatever the customer is willing to accept. The fewer questions customers ask, the greater the revenues for the carriers. The next six chapters will teach you how to proactively manage the costs of your local telephone bills.

5

Local lines

Some items on telephone bills are *fixed* charges, while others are *variable*. Line charges, for example, remain the same each month, but directory assistance charges vary each month according to usage. The more you call, the more it costs you. The first step in auditing your local bills is to understand the most common types of telephone lines and how they are billed. Plain old telephone service (POTS), trunks, and Centrex are the three most common types of local service. As a *tariffed* item, the cost for a phone line is nonnegotiable. However, line charges can still be reduced using one of the methods explained in this chapter. After a brief explanation of each type of local service, I will offer some proven cost-management strategies regarding local lines.

The physical telephone line

In spite of all the hype about fiber-optic phone lines, most businesses and residences still connect to the public-switched network using copper wires. Carriers are beefing up their networks by using high-capacity fiber-optic lines, but most of this is between central offices, which are commonly referred to as the "backbone" of the network. It is far too costly to run fiber-optic lines across the "last mile" to the end user. The copper wires are already in place, and it is just too expensive to replace the wiring. No matter how advanced the phone network, a user is still hindered by the limitations of the last mile. The last mile is almost always low-capacity copper wire. This trend is not likely to change any time soon, as the average cost for a "fiber drop" replacing the copper wires into a residence with fiber-optic lines costs approximately $2,000 to $3,000 at the time of this writing. Phone companies are too busy concentrating on competition and profits to make large investments upgrading the last mile from copper to fiber. Therefore, most of us have telephone service via traditional copper wires, rather than high-tech fiber-optic lines.

Circuit is a generic term referring to the physical connection, or path, between two given points, such as your phone and the central office. A common type of circuit, used in residences and businesses, is called "four-wire." One wire is used to transmit data, and one wire is used to receive data. The other pair of wires is a spare set. That is why the phone company technician only works inside the house if a person orders a second phone line to his or her house. The technician does not need to string a whole new pair of wires all the way to the house from the telephone pole.

The wires connecting to the central office may be identical, but they are called different names according to the type of connection and its purpose. Local service is provided three ways: POTS lines, trunks, and Centrex. Table 5.1 shows a comparison of these three types of phone lines.

Table 5.1   Local Line Comparison

Local Line	Monthly Cost	Line Ratio	Dialing	Equipment
POTS	$15–$20 residential $40 business	1 line per device	Normal	Single-line phones, fax machines, modems, key systems, and alarms
Trunks	$60	Many inside phones share a single trunk	Dial "9" for an outside call	PBX
Centrex	$20–$40	1 line per phone	Dial "9" for an outside call	Single-line phones and PBX*

* In some areas, it is illegal to use Centrex lines in conjunction with a PBX.

POTS

Regular phone lines are called "POTS lines," which stands for plain old telephone service. POTS lines are used with single-line termination points, such as a telephone, fax machine, alarm, or modem. For each phone or device, one phone line is required. Residential customers use POTS. Key systems also use POTS lines. Figure 5.1 shows how POTS lines work.

POTS lines typically cost around $40 per month for a business, while residential lines cost $15 to $25 per month. Although residential lines are more expensive for the phone companies to maintain, the end user pays less because residential service is subsidized by the phone companies through the higher rates that businesses pay for phone service.

Trunks

A *trunk* is a circuit that connects two *switching* devices. The lines connecting two telephone company central offices are called trunks, but for the end user, the term trunk is used to describe the line that connects a customer's private branch exchange (PBX) to the telephone company's central office. A PBX is a computer-driven phone system.

A PBX is a highly sophisticated telephone switch located at the customer's premise with an attendant console. The PBX connects to a group of trunks that are accessed by the caller first dialing "9" to make an outside call. Inside calling from one internal extension to another does not tie up a central office trunk, because the PBX switches the call; only inside wiring is used. Figure 5.2 shows how trunks work with a PBX.

If a business has a PBX, it must purchase trunk lines from the phone company. The physical circuits are still the same copper wires it would have

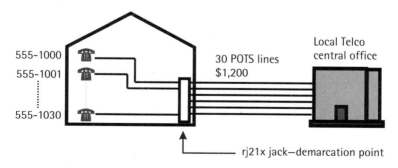

Figure 5.1 POTS.

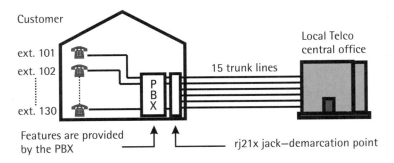

Figure 5.2 Trunk lines connect internal calls to the telephone company's central office.

had with POTS service, but with trunks, a business can expect to pay $60 per month, per trunk. The cost is higher than POTS lines, but multiple internal extensions can use one trunk, just not at the same time. Therefore, a business need only purchase a number of trunks equal to the number of calls it expects to have during its busiest time.

Centrex service

Centrex was originally designed for large customers, such as hospitals or college campuses, whose facilities were almost big enough to warrant their own central office or *central exchange*. The first Centrex system was implemented in 1965 at Prudential's office in Newark, New Jersey, but today, Centrex can be a cost-effective way to have sophisticated phone service even for a small business.

Centrex is a business service offered by local exchange carriers. These lines are normal telephone circuits with lots of features included in a package, such as call forwarding, call park, call transfer, speed dialing, and off-premise extensions. The local carrier's central office provides the features.

Like PBX service, Centrex service usually requires the caller to dial "9" to make an outside call (see Figure 5.3). This can be very cumbersome for the customer, especially when equipment may have to be programmed to dial the "9" such as alarm systems, computer modems, and fax machines. Some local carriers "assume the 9" so customers do not have to reprogram their equipment and retrain their employees.

Centrex pricing varies widely. Sometimes it costs less than POTS lines; sometimes it costs more. Centrex generally costs less than POTS lines with

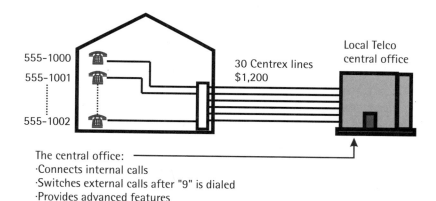

Figure 5.3   Centrex service.

GTE, U S West, Ameritech, and BellSouth. Centrex almost always costs more than POTS lines with Southwestern Bell. Centrex pricing with the other Bell companies varies.

Changing to Centrex service

Local carriers rolled out Centrex service in the 1970s as a way to offer advanced features to their customers. The customer got a reduced line charge but had to pay for the features. The total cost was usually higher than the cost of a single POTS line. Many customers figured out that they could replace their POTS lines with Centrex lines and save money. Even today, a customer can order "stripped down" Centrex lines without the expensive features. This can dramatically cut the cost of their local phone bill. Table 5.2 shows the cost savings of changing from POTS to Centrex.

Switching to Centrex requires a term agreement. In general, the longer the agreement, the lower the pricing. The pricing difference between a 5-year contract and a 1-year contract may only be 4% to 10%. Centrex volume commitments are more likely to specify a commitment of the number of lines rather than a dollar amount. BellSouth Centrex agreements usually specify that the customer must agree to keep a minimum of two lines during the term of the agreement.

The ideal candidate for Centrex is a business with numerous POTS lines and multiple features on each line because the features are free with Centrex. Some of the main cost-saving benefits of switching to Centrex are:

Table 5.2 Changing from POTS to Centrex

SERVICES		CURRENT PLAN—POTS	NEW PLAN—CENTREX
20 POTS lines	$40.00	$800.00	
Call waiting (5)	$3.50	$17.50	
Voice mail (5)	$5.00	$25.00	
20 PICC charges*	$2.75	$55.00	
20 Centrex lines	$28.50		$570.00
Call waiting (5)	Free		$ —
Voice mail (5)	Free		$ —
20 PICC charges	$0.31		$6.20
Total:		$897.50	$576.20
Monthly savings:			$321.30
Annual savings:			$3,855.60

*PICC charges are billed on the long-distance carrier's bill.

reduced line rates, free features, reduced local calling charges, and reduced primary interexchange carrier charge (PICC) fees. To find out if Centrex can help you save money, call your local telephone company and ask for a Centrex price quote.

Centrex line rates are often lower than POTS line rates. The main advantage to Centrex may be with the free features. A business that pays $5 per line on 10 lines for call forwarding each month should be able to save $50 a month by switching to Centrex and getting these features for free.

In many local markets, the local carrier's Centrex offering will include an option for an "off-premise extension." For example, a manufacturer with a factory and a sales office in the same city may be paying for local calls between locations. The company can convert the local service at both locations to be under one Centrex system through its local phone company. Then, instead of paying for calls between the two locations, the calls are free, because the lines at the sales office are now treated as off-premise extensions.

One of the so-called reforms of the Telecom Act of 1996 is that we all get to pay new fees, such as the PICC fee. For a long time, local carriers have charged customers a fee of $5 to switch long-distance carriers. This is called a change of the primary interexchange carrier (PIC). Recently, the local carriers argued that not only does it cost them to change a PIC code in their system, it also costs them to maintain the PIC code. The end result is that everyone now pays a PICC fee on each telephone line.

The LEC bills the long-distance carriers for the PICC, and they, in turn, bill you, the end user. The PICC fee shows up on the long-distance bill but is based on the type of service a customer has with the local carrier. Yes, it is confusing, but the good news is that Centrex customers pay a greatly reduced PICC fee. WorldCom has recently been charging a PICC of $4.31 for POTS lines, but only $0.31 for Centrex lines. Current Centrex customers are overbilled for the PICC almost a third of the time.

If you decide to switch to Centrex service, be careful to schedule the conversion during off-hours. It has been my experience that about a third of all Centrex cutovers result in a service outage for the customer. Some of the causes for the outage may be on the customer's premise, but sometimes the telephone company technician at the central office makes a mistake. Regardless of who is at fault, the business suffers the down time. It may be advantageous to hire a consultant, or deal directly with a Centrex agent who is skilled at conversions instead of dealing directly with phone company representatives.

Alternative local lines offered by local carriers

Most phone calls by consumers or businesses are carried across ordinary POTS lines, trunks, or Centrex lines. However, local carriers do have alternative offerings such as WATS, T-1, T-3, ISDN, dedicated private lines, DSL, X.25, frame relay, and tie lines. These services are covered in Chapter 16, but I will explain how some of these services can be beneficial for voice calls in this section.

WATS

Wide-area telecommunications service (WATS) was originally offered by AT&T prior to divestiture. It was designed for high-volume long-distance users and was more of a bulk pricing service than an alternative technology. It came in two varieties: In-WATS, the precursor to toll-free calling, and Out-WATS.

Long-distance carriers have moved their customers away from WATS service, but the local carriers are less proactive. A customer who used

WATS service for intralata calling 10 years ago might still have monthly billing for the service. A good portion of the WATS service today is inactive. Customers now use direct dial long distance and 800 services from their long-distance provider. If customers fail to cancel the WATS service, even though it is not being used, they may still be paying a hefty recurring fee of up to $500 each month for it.

Cancel WATS service

The reason this antiquated service is addressed in this book is because one in ten large businesses still has a WATS line. WATS lines were originally a great idea because they lowered a customer's cost for long-distance calling. With the decline of long-distance pricing over the past decade, most WATS lines should be canceled and regular long-distance calling should be used. The cost of direct dial long distance today is much lower than WATS pricing.

T-1 for local service

Local carriers have struggled to increase their network capacity to keep up with ever-increasing call volumes. The demand for lines is also rapidly increasing. Carriers are now using technologies that allow more phone calls to travel across the same number of phone lines. T-1 service accomplishes this objective by using *multiplexing* technology, which is a method of cramming multiple phone calls across the same phone line at the same time.

A T-1 is a *dedicated* connection between the customer's premise and the carrier's central office. A T-1 has 24 distinct channels for calls. It can carry 24 voice or data conversations at the same time. The 24 channels are multiplexed across four wires (which is the same physical capacity as two POTS lines). Because less of the carrier's network is used, local carriers are able to offer high-bandwidth T-1 service at a fraction of the cost of the same bandwidth across multiple POTS lines. Figure 5.4 shows how multiplexing works.

Saving money with a local T-1

Because the T-1 can handle 24 simultaneous calls, it can theoretically be used to replace 24 local trunks or POTS lines. PBX trunks typically cost $60 per month, but T-1 service is available in most markets for less than $500 per month. (Exact T-1 pricing is determined by the actual mileage from the central office to the customer's premise.) Using these figures as a guide, the

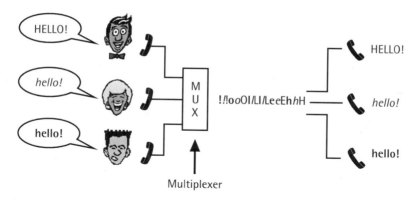

Multiplexer

Figure 5.4 Multiplexing technology.

break-even point is about 10 phone lines. Factoring the "pain of change" into the equation, a business with 15 or more trunks should definitely consider switching to T-1 service.

Table 5.3 shows the monthly savings a business can expect by replacing 20 trunks with one T-1. Note the change in billing due to lowered PICC charges and FCC line charges.

Table 5.3   Replacing 20 Trunks with a Single T-1

Services		Current Plan—Trunks	New Plan—T-1 for Local Service
20 trunks	$60.00	$1,200.00	
20 FCC charges*	$8.13	$162.60	
20 PICC charges**	$2.75	$55.00	
T-1			$500.00
1 FCC charge	$8.13		$8.13
1 PICC charge	$2.75		$2.75
Total:		$1,417.60	$510.88
Monthly savings:			$906.72
Annual savings:			$10,880.64

*FCC charges are billed on the local bill.
**PICC charges are billed on the long-distance carrier's bill.

These monthly savings are significant, but this is an often-overlooked cost management measure. An organization's financial managers need technical expertise to handle the T-1 conversion. The company's technical managers are often too busy with new technologies to be concerned about upgrading a low-tech service such as trunk lines, especially if the current service is functioning properly. Once you have done your own T-1 cost study and determined that it is beneficial, then it is time to pursue a technician, such as your PBX maintenance company. The technicians can tell you exactly what needs to be done to your equipment to upgrade to T-1 service. (T-1 service is explained in greater detail in Chapter 16.)

How line charges and local calls are billed

Regardless of whether or not a business has lines, trunks, or Centrex, there are three ways to pay for local phone lines. With *flat-rate* service, the line is billed but local calls are free. *Measured* service consists of a line charge, and local calls are billed by the minute. *Message-rate* service has a line charge, and local calling is billed on a per-call basis. Each call is considered a "message." Table 5.4 shows a comparison of these three classes of local service.

Reducing line charges by changing class of service

Local telephone companies usually only offer one or two of these classes of service in each market. The best practice is to compare your current service to whatever alternative service the carrier offers. If you change your class of service, it is usually only a billing/pricing change—not a change regarding how the service actually works.

Line charges are usually higher with flat-rate service. A business such as a mail-order company that makes few local calls can benefit by changing

Table 5.4   Three Different Classes of Service for Local Lines

CLASS OF SERVICE	LINE RATE	COST OF LOCAL CALLS
Flat-rate	$40.00	Free
Measured-rate	$30.00	$0.03–$0.04 per call
Message-rate	$30.00	$0.01–$0.02 per minute

from flat-rate service to message-rate. The line charges will be lower, and because the business makes almost no local calls, the local calling charges will be minimal. Table 6.2 shows an example of this type of change. The math, of course, works two ways; a business that makes a great deal of local calls will profit by switching from measured-rate service to flat-rate service.

Saving money by changing to a CLEC

CLECs are aggressively winning business away from incumbent local carriers. CLECs usually enter a local market as a reseller of the incumbent LEC's services. Once they have established a customer base, they will begin to install their own lines and central office switches. Teligent, one of the first CLECs, offered local service via line-of-sight wireless connections. (Building a network from the ground up is costly, and cash-poor Teligent went bankrupt in May 2001.)

As part of the Telecom Act of 1996, incumbent LECs are required to allow CLECs to interconnect to their network. They are also required to allow the CLECs to "colocate" their switching equipment in the same central office.

CLECs usually offer prices 5% to 30% lower than the incumbent LECs. The end user has to decide whether to sign up with the CLEC under a *reseller* arrangement or under a *facilities-based* arrangement. The savings are better using the CLEC's actual facilities, but many of the problems previously mentioned regarding a Centrex conversion apply with a change to a CLEC's physical lines. It may not be worth the "pain of change" for the customer, so the reseller option may be best.

Eliminating erroneously billed lines

Many businesses make a "donation" to their local telephone company each month by paying for lines that are not used. Once a year, a business should audit its phone records for these lines. A low-tech method of verifying what lines you have is to request a list of all your telephone numbers from the local phone company, and call the numbers yourself. You may be surprised to see who answers.

In some cases, a business may be paying for someone else's lines because the telephone company made an error in its records. A simple data entry error may cause you to pay for your neighbor's phone lines. Or you might be paying the phone bills of an ex-employee's home office phone bills each month. Unless you are feeling benevolent, you should correct the problem. Notify your local carrier and request a refund of all the past-due charges plus interest. The carrier will reroute the phone bills, but the actual customer probably will not be back-billed.

Reducing the number of trunks

As noted above, a business with a PBX uses trunk lines provided by the local telephone company. The trunks connect the customer's PBX to the telephone company's central office. Within the customer's premise, inside wiring connects each individual phone to the PBX. This configuration is often called "trunks and stations."

A business only needs a number of trunks equal to the maximum number of callers who will be making outside calls at the same time. To determine whether or not you are "overtrunked," you can order a trunk study from your local carrier. A *trunk study* measures the number of outgoing calls over a specified period of time. The term *traffic study* is used to describe a usage study performed on POTS or Centrex lines.

To order a trunk or traffic study, call your local telephone company representative and specify when you want the company to analyze your traffic volume. Your carrier will probably give you a 5-day statistical sampling of your traffic volume. The trunk study report sent to you details the number of calls, the duration of the calls, the "busy hour" each day, and the amount of any "overflow."

In the sample trunk study in Figure 5.5, the customer experiences "call blockage" on Thursday and is, therefore, "undertrunked." This study reveals the need to add, not drop, trunks. This is especially important for a business that cannot afford to miss any calls, such as a hospital. A radio station that experiences high levels of overflow should not be concerned, because the overflow is most likely callers flooding the lines in response to an on-air contest.

Miscellaneous recurring charges billed by the LEC

In addition to line charges, local calling, and intralata calling, the local phone bill will include additional charges. Most of these items are optional but some are almost always mandatory, such as the FCC charge and touch-tone service. Optional services are normally nonregulated, so charges will vary from market to market. One thing these services have in common is that they are services of convenience. For example, the average

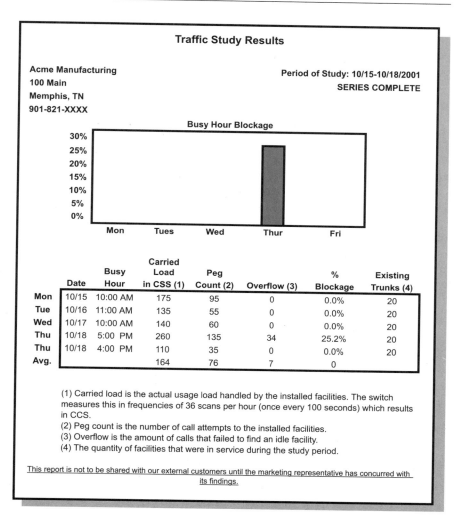

Traffic Study Results

Acme Manufacturing
100 Main
Memphis, TN
901-821-XXXX

Period of Study: 10/15-10/18/2001
SERIES COMPLETE

Busy Hour Blockage

	Date	Busy Hour	Carried Load in CSS (1)	Peg Count (2)	Overflow (3)	% Blockage	Existing Trunks (4)
Mon	10/15	10:00 AM	175	95	0	0.0%	20
Tue	10/16	11:00 AM	135	55	0	0.0%	20
Wed	10/17	10:00 AM	140	60	0	0.0%	20
Thu	10/18	5:00 PM	260	135	34	25.2%	20
Thu	10/18	4:00 PM	110	35	0	0.0%	20
Avg.			164	76	7	0	

(1) Carried load is the actual usage load handled by the installed facilities. The switch measures this in frequencies of 36 scans per hour (once every 100 seconds) which results in CCS.
(2) Peg count is the number of call attempts to the installed facilities.
(3) Overflow is the amount of calls that failed to find an idle facility.
(4) The quantity of facilities that were in service during the study period.

This report is not to be shared with our external customers until the marketing representative has concurred with its findings.

Figure 5.5 Trunk study report.

business can usually live without call forwarding or voice mail, but these services certainly make doing business more convenient.

Optional services are usually listed in the first pages of your local phone book. Carriers change their service offerings frequently, so you may want to call the carrier to find out exactly what is being offered today.

Does the money for the "FCC Charge" really go to the FCC?

In addition to the charge for the actual phone line, local carriers also charge a fee with each line. During the negotiations associated with the breakup of AT&T in the early 1980s, the Bell companies argued that AT&T would get more money because long-distance service has higher profit margins than local service. They also complained that they had to bear the high cost of maintaining the local lines from the central office all the way to the business or residence. This was a legitimate concern, so the government allowed the LECs to subsidize their costs by charging each customer a fee per line.

On the phone bill, this fee usually reads "FCC Approved Line Charge" or "Interstate Access Fee." The fee is usually around $8 per line. The money does not actually go to the FCC—that organization is funded by our taxes. Instead, this money goes directly to the local carrier as additional revenue. Over the years, this fee has steadily increased, like most other charges on the local phone bill.

This fee is not negotiable, so the chances of saving any money here are slim. The only way to cut this cost is to order your customer service records from the local phone company and ensure that the number of fees charged does not exceed the number of lines charged. (Customer service records are covered in detail in Chapter 10.)

Optional services

Most optional services are nonregulated, so the charges vary from market to market. One thing these services have in common is that they are "services of convenience." For example, the average business can usually live without *call forwarding* or *voice mail*, but these services certainly make doing business more convenient. Some of the most common optional services are:

- Remote call forwarding;
- Wire maintenance;
- Voice mail;
- Pay-per-use features.

Remote call forwarding

Remote call forwarding (RCF) allows a subscriber to have a local number in one area from which all calls are forwarded to another area, such as the home office. For example, Pete's Plumbing has an office in Orlando, but Pete and his plumbers work across Florida. When someone in Miami needs a plumber, they look in the Miami phone book and call a local plumber. Pete is not in Miami, but RCF gives the impression that he is in Miami. Pete pays $15 per month for RCF so he can have a Miami number. When the customer in Miami calls Pete's Plumbing, the call is forwarded back to Orlando. Unless Pete tells him otherwise, the customer in Miami thinks the company is a local business. Pete incurs long-distance charges on these calls, but the additional customers make it worthwhile.

Saving money on RCF RCF is a tariffed item, so the rates are set. About the only way to save money on this item is to cancel the service altogether and replace it with 800 numbers. Compare all of the costs, including 800 number fees, long-distance rates, and call volume, before switching. The main advantage of remote call forwarding is the perception it creates that the company is a "local business."

Wire maintenance

The urban legend states that the CEO of one of the Regional Bell Operating Companies challenged his managers to "find a way to generate an additional $2 million in annual revenue, without increasing the level of service we provide our customers." The end result was the wire maintenance plan that was quickly adopted by local carriers nationwide. Some of the names of this plan used by the different carriers are: Linebacker, Lineskeeper, Inside Wiring Plan, Trouble Isolation Plan, and Wire Keeper.

The local telephone company is not responsible for maintaining the physical wires inside a customer's premise beyond the demarcation point. The wire maintenance plans work like an insurance policy. A customer usually pays approximately $3 per line per month for the wire maintenance plan. In the event the inside wiring fails, the phone company technician will repair the inside wiring. The main problem with this plan is that inside wiring almost never fails.

Another problem with wire maintenance plans is they often do not cover the entire cost of the technician's repair job. A typical wiring repair job carries these charges:

- ◆ Initial trip to the customer's premise;

- ◆ Cost to diagnose the problem;

- Cost to repair the problem;

- Cost of materials.

In the case of Ameritech's $3.50 Wire Maintenance Plan, only the diagnostic charge is covered. To have the line repair covered, the Linebacker plan must be added for an additional $3.50 per month. To be "fully insured," a customer pays $7 per month to be protected from a $50 to $100 repair that only occurs once every 5 to 20 years.

The biggest sting of wire maintenance plans is that the average customer is unaware that it even has the plan to begin with. A manufacturer in the Chicago area was donating more than $100 each month to Ameritech, because it had the wire maintenance plan on more than 50 lines. In many cases, the wire maintenance plans were added to customer bills without their permission. Class-action lawsuits against the RBOCs have recently been resolved. The RBOCs had to give back millions of dollars to their customers who had wire maintenance service.

Most businesses have their phone bills managed by the accounts payable department, but the phone service itself is managed by the computer department or telecom department. In the event the lines need repair, the telecom department will probably call its phone equipment vendor to do the work, and not the local carrier.

Cancel wire maintenance plans   Because of the reasons listed, almost every business should cancel its wire maintenance plans. When canceling the plan with your local carrier, request a full refund of back charges unless your organization actually did order the service. Sometimes, the local carrier will issue a 3-month "courtesy credit" anyway.

Voice Mail

On numerous local bills, there is a $5 to $10 charge for voice mail. A decade ago, this may have been a cost-effective way to have voice messaging. Many Centrex services include voice mail as part of the package of additional services.

Saving money on voice mail   If you see voice mail being charged on your local bill, first find out if it is even being used. Your staff should be able to tell you who is using the service. If not, check with the phone company to see who set it up. In many cases, these "mystery services" were set up accidentally by an overzealous phone company representative or by an ex-employee. In either case, canceling the service can reduce your bill.

If you have a large number of voice mailboxes, you may be able to negotiate a lower price with the telephone company, or you may want to look into switching to a company that specializes in offering voice mail separately. Another cost saving idea is to buy your own voice mail equipment. If you have a PBX, it may be as simple as buying a voice mail card and installing it in your PBX.

Pay-per-use features

Pay-per-use features have been described as a lazy man's way of using the phone. Every time the caller uses the service, a charge is generated, thus the term *pay-per-use*. The most common charges are directory assistance call completion, call return (*69), repeat dialing, and three-way calling. The charge is typically $0.75 per use.

Saving money on pay-per-use features   Employees typically do not worry about the small cost associated with these charges, but they can potentially be a significant expense for a business. I have seen businesses pay thousands of dollars per year for the convenience of using these unnecessary services. The simplest way to cut your cost here is to call the local carrier and have it block these features in its central office. If you decide these services are necessary, consider purchasing one of the bulk packages offered by the local carriers.

Directory services

Phone books contain two types of listings: white pages and yellow pages. Each separately billed local account is entitled to one white page listing. Enhanced white page listings that include bold text or additional lines cost extra. Having your numbers *unlisted* will exclude them from the directory but they will still be available through directory assistance operators. A *nonpublished* number is not printed and is not available to operators.

Yellow pages listings are usually sold once a year, although the billing shows up at the back of the local phone bill each month. In some cases, the billing is a one-time expense. Pricing is based on the size and placement of the advertisement, if any, and the number of listings.

Saving money on directory services   Directory services are nonregulated, so the rates have a greater propensity for error. One of the most common errors is that a business may have some of its numbers incorrectly designated as "nonpublished" and pay $3.50 per month for this. I have seen payphones in a warehouse treated as "nonpublished." This is ridiculous because the local carrier typically does not publish payphone numbers in its directory anyway.

The most significant errors to occur with directory services are with the yellow pages. When an order is placed for the advertisement, the phone book is printed as far out as 1 year later. Sometimes the advertisement in the book does not match what is ordered. In this case, the business is entitled to some form of refund. Often, the yellow pages company (usually a sister company of the LEC) offers reduced rates on the advertisement in next year's phone book. If it misprints the number, you can have the company pay to secure the erroneous number and call forward it to your correct number. If the erroneous number is not available, you may be out of luck.

One of the most recent scams in the yellow pages industry has to do with fraudulent billing by other companies. Companies have sprung up that print their own phone books to compete with the standard phone book provided by the local carrier. Usually, the books are legitimate (although not as thorough as the "real" phone books), but their tactics are unscrupulous at times.

When securing a new order for a directory advertisement, sometimes these companies simply mail an invoice to a business in hopes that the company pays it. These invoices look genuine because they include the yellow pages logo, which is not a trademarked logo. The accounts payable clerk who processes the bills understands that yellow page billing is a legitimate charge, so the invoice gets paid. The "dummy invoice" is a fraudulent technique becoming increasingly common, and not just with yellow pages, or telecommunications, for that matter.

6

Local calls

Every phone call is a local call, an intralata call, or a long-distance call. These three classifications are determined by geographic boundaries. Figure 6.1 shows the two LATAs in Washington state. A call from Yakima to Seattle crosses LATA boundaries and is therefore a long-distance call. A call from Yakima to Spokane is within the LATA and is therefore an intralata call. A call made within Yakima is a local call.

Local calling boundaries are set by the local carrier; LATA boundaries were established at the time of divestiture. To know the exact boundaries of your local calling area, consult the map in the front of your local phone book. Most consumers do not pay for their local calls—they have *flat-rate* local service. They pay for the line, but the local calls are free.

Businesses either have flat-rate, measured-rate, or message-rate service. Customers with measured-rate service pay for their local calls according to the minutes of use. Message-rate customers pay for local calls according to the number of messages (calls). Table 6.1 shows a cost comparison of these three types of service. With all three classes of local service, no one pays for incoming local calls.

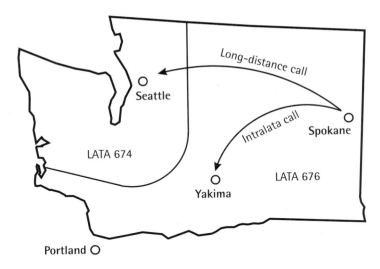

Figure 6.1 Local access and transport areas in Washington state.

Table 6.1 Class of Service for Local Lines

CLASS OF SERVICE	MONTHLY LINE CHARGE	COST OF A LOCAL CALL
Flat rate	$40.00	Free
Measured rate	$30.00	$0.02 per minute
Message rate	$30.00	$0.08 per call

Saving money on local calls by changing class of service

Reviewing the class of service for your local lines is an integral part of a telecom audit. Businesses can change from flat-rate service to measured- or message-rate service. Certain types of business are key candidates to change their service.

For example, a telemarketing company whose calls are all long distance should switch from flat-rate service to measured-rate service. Its line charges will be lower with measured-rate service and, because the telemarketing company makes minimal local calls, the charge for usage will be lower. Flat-rate service is better for a business with a lot of local calling.

Table 6.2 shows the cost comparison of a telemarketing company that switched from flat-rate to measured-rate local service. The telemarketing company has 50 lines, and each line averages only 60 minutes of local calling each month. In this example, the customer can save more than $5,000 per year by making this change.

This strategy does have a caveat. The phone bill for flat-rate local service does not give a summary of the local calls. Therefore, the customer does not know exactly how much local calling it has done. I have seen one business switch from flat-rate to measured-rate service only to find its costs increase by $800 per month. The company was unaware that its local calling was so high. Unless you are sure your local call volume is low, you should first find out exactly how much local calling you have. This can be accomplished by having your local carrier perform a traffic study. Your telephone equipment may also be capable of tracking the volume.

When considering changing the class of service for your local lines to reduce the cost of your local calls, the following items must be included in your cost comparison: number of local calls, duration of local calls, and the difference in the cost of the lines.

Off-peak calling

A low-tech way to cut the cost of your local calling is to use lower off-peak calling rates. This works especially well if you have a great deal of computer modem traffic. If you do not use dedicated data lines, then your computer modem calls are billed as regular voice calls. If you have a significant amount of modem traffic, consider changing the time that you transmit the data to be evening off-peak calling. Phone companies offer reduced rates at night to encourage callers not to flood the network during peak hours.

Table 6.2   Changing from Flat-Rate to Measured-Rate Local Service

Flat-Rate Service	Measured-Rate Service
50 local lines × $40.00 = $2,000.00	50 local lines × $30.00 = $1,500.00
3,000 local calling minutes = Free	3,000 local calling minutes × $0.02 = $60
Total: $2,000.00	$1,560.00
Monthly savings:	$440.00
Annual savings:	$5,280.00

Local calling packages

As in the case with Bell Atlantic's ValuePak calling plans, some local carriers offer discount local calling plans. A good rule to follow when auditing local bills is that anytime you see calls billed according to measured usage, there may be a less expensive way to have the calls billed. A quick phone call to the carrier is all you need to find out what options you have. If the carrier has no discount plans for local calls, consider eliminating the calls altogether by switching to flat-rate service.

Some local calling plans offer a discount each month, such as 10% off, while other plans simply offer lower per-minute calling rates. The impact of the plan is usually fairly simple, but the plan's design may be puzzling. Consider Bell Atlantic's ValuePak plan that has been offered for years in Pennsylvania. The plan allows you to purchase "Paks" of local calling at a reduced rate each month. For $13.80, you get a calling allowance of $18. If you use more than your allowance, the rest is billed without a discount. If you use less than your allowance, you are still billed $13.80 per month. Higher volume users usually add multiple ValuePaks but are limited to one Pak per line on the account. Table 6.3 compares a customer's $100 cost with and without ValuePaks.

As illustrated in Table 6.3, the customer can save $21 per month by adding five ValuePaks. I have also encountered numerous customers that pay for ValuePaks but have no local calling. Either their calling habits changed or they moved their local calling traffic to another carrier. If they do not cancel the ValuePaks, they will continue to make a donation to Bell Atlantic every month.

Table 6.3   Bell Atlantic's ValuePaks

COST WITHOUT VALUEPAKS		COST WITH 5 VALUEPAKS		
Local calling	$100.00	Local calling	$100.00	
		5 ValuePaks × $13.80 =		$69.00
		Calling allowance	$90.00	
		Additional cost	$10.00	$10.00
Total:	$100.00	Total:		$79.00
Monthly savings:				$21.00
Annual savings:				$252.00

Using local call volume to secure discounts

Some discount plans offered by LECs do not offer lower rates for local calls, but they do allow the local calling volume to contribute to the overall volume. As with most telecommunications pricing, the greater your volume, the greater your discounts.

For example, a customer who signs up for Ameritech's Complete Link plan with a 12-month term and a $500 monthly volume commitment receives lower rates for intralata calling and a discount of 4% to 5%. The discount is applied to monthly service, local calls, and intralata calls. While the local calls do not directly receive a reduced rate, the local calling volume may be significant enough to allow the customer to qualify for the next discount tier. Table 6.4 shows an example of the Complete Link plan.

Shifting local calls to an alternate carrier

In some markets, the local provider does not offer low rates for local calls. A business in this situation may be able to cut its costs by switching to an alternate carrier for its local calls. Changing your local calling provider is not as simple as changing your long-distance provider, but it may be worthwhile if you can save enough money.

Using the long-distance carrier's T-1 for local calls

Local calls are normally carried across LEC trunks to the central office and then the call is connected to its destination (see Figure 6.2). But if you have

Table 6.4   Ameritech's Complete Link Plan

WITHOUT COMPLETE LINK		WITH COMPLETE LINK	
15 local lines × $38.50 =	$577.50	15 local lines × $38.50 =	$577.50
2,500 local calling minutes × $0.02 =	$50.00	2,500 local calling minutes × $0.02 =	$50.00
3,500 intralata minutes × $0.15 per minute =	$525.00	3,500 intralata minutes × $0.08 per minute =	$280.00
		5% complete link discount	$45.38
Total:	$1,152.50		$862.13
Monthly savings:			$290.38
Annual savings:			$3,484.50

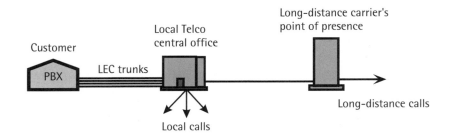

Figure 6.2 Local calls originate at the customer's premise and travel across local telephone company trunks.

dedicated service through your long-distance provider, you can easily switch your local calls to your long-distance provider from your local carrier. Dedicated service means you have a T-1 connection from your facility directly to the long-distance carrier's central office (see Figure 6.3). The outbound local calls can be rerouted away from the local provider by reprogramming the PBX to handle local calls as long-distance calls. Incoming local calls will still use the local carrier's trunk lines.

In a typical example, the PBX sends these local calls to AT&T instead of the incumbent local provider, such as Pacific Bell. AT&T has been calling this service Digitalink. By reducing the cost of their local calls from $0.02 per minute to $0.01 per minute, I have seen many Digitalink customers save more than $500 each month.

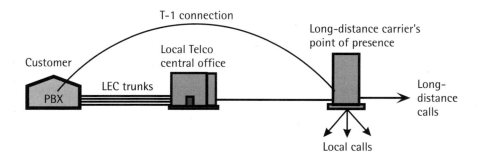

Figure 6.3 Local calls originate at the customer's premise and travel across the T-1 connection.

CLECs

The second way to move your local calls away from your current provider is to switch carriers altogether. The new CLECs would be more than happy to have your business. A business with a significant amount of expensive local calling whose local service is not complex is a good candidate to switch to one of the many new CLECs. Be careful that they do not also secure your long distance, however, unless you want them to.

OPX: The off-premise extension

A business with two locations in the same city is billed for local calls between the two facilities. If it has Centrex service, however, it can eliminate these charges. With Centrex, one location can be designated as an off-premise extension (OPX).

The OPX functions like an internal extension, and calls between the two facilities are handled like internal calls rather than local calls. Converting an off-site location to an OPX eliminates all local calling charges between the two locations. A more detailed explanation of Centrex is given in Chapter 5.

Zone calls

In certain markets, customers are billed for *zone* calls, in addition to the local calls and the intralata calls. This is true for Ameritech's Detroit, Michigan, customers. As Figure 6.4 reveals, the zone is a geographic area in-between the local area and intralata area.

Zone calls are like local calls in that the phone bill gives no detail for these calls. Zone calls are like intralata calls, in that zone calling rates are much higher than local calling rates. But unlike local calls and intralata calls, the cost of zone calling is difficult to reduce. However, you can have your zone calling volume contribute toward a greater volume discount with your carrier.

Ameritech's Value Link plan offers lower intralata rates based on a term and volume commitment. Although the zone calls are not directly reduced, at least the volume from these calls contributes to the overall Value Link volume commitment. The Value Link plan is being replaced by the newer Complete Link plan.

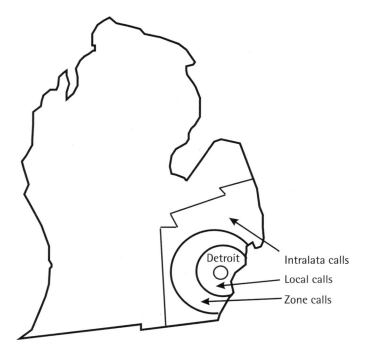

Figure 6.4 Zone calls in Detroit, Michigan.

7

Intralata calling

LATAs were set up as a result of the landmark breakup of AT&T called divestiture. Divestiture was implemented on January 1, 1984. The LATA is a geographic area wherein the local carrier, such as U S West, provides all the land line-based telecommunications services. *Access* refers to the customer's connection to the public-switched network. The phone line that connects to U S West's central office gives a customer "access" to the outside world. Within the LATA, U S West is designated to carry, or *transport*, all of the calls.

Figure 7.1 demonstrates the concepts of access and transport. Ron "rents" a phone line each month so he can have "access" to his friends via the telephone. When Ron dials the number, the local telephone company "transports" the call to Bev's phone. A call originating and terminating within the same LATA is called an intralata call. Depending on the carrier, intralata calls may show up on local phone bills under any of the following headings:

- Local toll calls;
- Local calls;

Figure 7.1 Since divestiture, local telephone companies have provided access and transport for all calls within the LATA.

- Long-distance calls;

- Itemized calls;

- Regional toll calls;

- Toll calls;

- Measured calls.

Even though long-distance rates dropped significantly in the decade after divestiture, intralata call rates remained fairly level during this time. It was not uncommon for a business in Dallas, Texas, to pay $0.15 per minute to call Ft. Wayne, Indiana, but pay $0.25 per minute to call to Ft. Worth, Texas. Customers thought call rates should be lower if the person called is only a short distance away. Due to a lack of competition for intralata calling, local carriers did not reduce their rates. Customers began to look elsewhere for service, and the first place they looked was their long-distance carriers.

Autodialers

Long-distance carriers knew that if a customer dialed its carrier identification code (CIC), such as 10-10-288 for AT&T today, the call would be routed to AT&T. This works regardless of whether or not AT&T is the chosen long-distance carrier, and it works regardless of whether or not the call is intralata or interlata.

Because the long-distance carriers knew that the average business would not take the time to dial the CIC, they installed *autodialers*, which are small electronic devices that interpret the numbers that the caller is dialing. If the autodialer detects the dialing of "1 + area code + number," then it automatically inserts the CIC before the phone number the caller has dialed. When the local carrier's central office computer detects the CIC; it automatically sends the call to the long-distance carrier's network.

With this fairly simple technique, long-distance carriers were able to capture the intralata traffic from the local carrier and increase their revenues. The cost of a dialer is around $300, and as long as a customer used $50 or more in monthly intralata traffic, most carriers installed autodialers at the customer's premise for free. The caveat of autodialers is that they have a reputation of being electrically unstable. The slightest power surge due to a nearby lightning strike is all it takes to reset the average autodialer. When that happens, your technician must reprogram the equipment. Keep that number handy for the next time a thunderstorm rolls around.

Dial-around

The easiest way to shift your intralata traffic from your local carrier to your long-distance carrier is to use the newly popular dial-around techniques. This works the same way autodialers work, except the caller is manually dialing the CIC instead of the autodialer. This solution never works with a larger company, because the cost of training the staff outweighs the savings. For a small office with few employees, however, this can be a very effective way to trim intralata billing.

Program your PBX

Another technique for moving intralata traffic to your long-distance carrier is to program your PBX to function like an autodialer. The PBX can insert your carrier's CIC and move all intralata calling away from the local carrier. If you have a key system, or single-line phone set, you should be able to program one of the speed dial buttons to dial the CIC for you.

PIC-ing your intralata traffic

In many ways, the Telecommunications Act of 1996 has reversed some of the changes implemented at divestiture. Divestiture was designed to break up AT&T's monopoly and allow for a more competitive marketplace. At divestiture, intralata traffic was ruled to be carried only by the local carrier. The Telecommunications Act of 1996 was designed to open up local markets for more competition, and now long-distance carriers can carry intralata traffic again. In order for local carriers to be allowed into the long-distance market, they must first open up their market to competition. Intralata calling is the first area local carriers opened up for competition.

Prior to 1996, customers had to use one of the techniques already listed to move their intralata traffic to their long-distance provider. Today, in most markets, intralata traffic is treated like long-distance traffic. Just like a customer's choice of long-distance carriers is maintained in the LEC's central office, the same holds true for a customer's carrier choice for calls within the LATA. To change your intralata carrier, simply call your local phone company. If it confirms that it has opened up this part of its market for competition, the phone company can change your intralata carrier immediately.

Do not be surprised if the company tries to persuade you to keep the traffic with it. If it has a calling plan, the local carrier will offer it at this time. Intralata traffic brings in a lot of revenue for local carriers, and they hate to lose it. This is a good opportunity to get your two carriers into a bidding war for your business. Just do not expect them to pursue you unless you have more than $500 in monthly intralata call volume.

LEC calling plans

Prior to moving your intralata traffic away from your local carrier to your long-distance carrier, ask the local carrier first if it has any discount plans available. Being very mindful of the new competition for this traffic, the LECs are rolling out new competitive offers regularly. Some of the recently offered intralata calling plans are:

Ameritech: Valuelink, Complete Link;

Bell Atlantic: Key Connection;

BellSouth: WatsSaver;

Pacific Bell: Value Promise Plus;

SBC: 1+ Saver;

U S West: Calling Connection Plan;

GTE: Easy Savings Plan, Platinum Plan.

The intralata calling plans offered by the local carriers are usually based on a volume and term commitment. The long-distance carriers rarely require a commitment. They are happy to give you aggressive rates just to secure the added call volume to their network.

When considering an intralata calling plan *term* option, compare the rates carefully. The terms offered are usually 12-, 24-, and 36-month options. The carriers will push you to sign a 36-month agreement, but the pricing difference between the 36- and 12-month options is often very minor. Too often, a customer signs a long-term agreement with a phone company, only to find that halfway through the agreement, the company's pricing is no longer competitive. In most cases, the customer is stuck with the original agreement. Carriers are only willing to cancel an existing term agreement if they replace it with another one.

6-second billing

Another relevant item to consider at this point is the *billing increment* used by the carrier. Since divestiture, most local carriers have billed their intralata traffic in full-minute increments. Each call is rounded up to the next whole minute. For example, a 66-second call is billed as a 2-minute call. Long-distance carriers typically bill in 6-second increments. The 66-second call would therefore be billed as a 1.1-minute call. With most telephone calls only lasting about 3 minutes, this incremental billing issue is a significant concern.

Table 7.1 illustrates the value of 6-second billing over full-minute billing. In this example, the customer can trim its bill by 8% by changing to 6-second billing increments. Some carriers, such as Qwest, use a 1-second billing increment, which can reduce a customer's bill even more.

Courtesy credits

For the past 10 years, local carriers have been charging rates as high as $0.35 per minute for intralata calls. In some cases, it may have been cheaper for the caller to drive his car across the LATA boundary and have a face-to-face conversation. When it costs more to call 10 miles away than it does to call Japan, something is wrong with the system.

Table 7.1   Subtle Overbilling of Full-Minute Increments

ACTUAL CALL TIME	FULL-MINUTE BILLING INCREMENTS		6-SECOND BILLING INCREMENTS	
(Minutes: Seconds)	Billed Time	Actual Cost	Billed Time	Actual Cost
0:30	1.0	$0.10	0.5	$0.05
1:00	1.0	$0.10	1.0	$0.10
1:30	2.0	$0.20	1.5	$0.15
2:00	2.0	$0.20	2.0	$0.20
2:30	3.0	$0.30	2.5	$0.25
3:00	3.0	$0.30	3.0	$0.30
3:30	4.0	$0.40	3.5	$0.35
4:00	4.0	$0.40	4.0	$0.40
4:30	5.0	$0.50	4.5	$0.45
5:00	5.0	$0.50	5.0	$0.50
Total:		$3.00		$2.75
Cost difference:				$0.25
% difference:				8%

Local carriers are aware of this pricing discrepancy and are often willing to offer a one-time courtesy credit, or loyalty credit, for signing up for a new calling plan with them. Of course, they are not required to give credits, but if you have been paying more than $0.15 per minute for intralata calls, you have been grossly overcharged, and you deserve a refund. In most cases, to satisfy a customer, service representatives can give a credit of up to $500 without their manager's approval.

Extended local calling area plans

As previously noted, local carriers bill both local calls and intralata calls. Local calls are normally only a few pennies per minute, while intralata calls cost $0.05 to $0.15 per minute. In some markets, the local carrier will allow you to pay to have your local calling area enlarged.

By extending your local calling area, some intralata calls will now be treated as local calls. In the examples in Figures 7.2 and 7.3, the customer paid $0.15 per minute to call Pleasantville. But after implementing the extended local calling area plan, these calls are now billed as local calls and only cost $0.02 per minute.

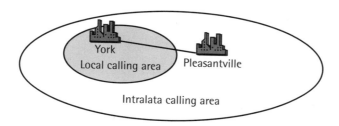

Figure 7.2 A typical local calling area.

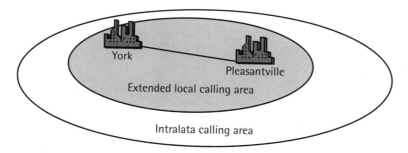

Figure 7.3 An extended local calling area.

8

Long-distance calls on local bills

During the past few years, local phone bills have become like credit card accounts. A host of services can be billed on the last pages of your local phone bill, including charges for long distance, 900 calls, collect calls, Internet charges, and miscellaneous fees. Sometimes these charges are legitimate—but often they are fraudulent; and they are always expensive. Regardless of why the charges were billed, a business can always reduce this expense, or eliminate it altogether.

Local telephone companies allow other carriers to tack their charges onto your local bill because they keep a percentage of the charges. Other companies do not mind paying this commission because local carriers collect the money for them. It is a win-win situation for both companies, but not for the customer. The average customer does not question any charges on the local bill. She sees it as an "assumed cost" and pays the bill each month. Even if a customer suspects that the bill is incorrect, he will still pay it, rather than risk having his local service disconnected. In truth, however, local carriers will not disconnect a customer's service for withholding payment for another company's charges.

Loose traffic

Loose traffic is a term widely used by AT&T referring to long-distance traffic billing on a local bill, instead of on the master long-distance account. Loose traffic usually bills on the back pages of the local bill or on a separate long-distance bill. Other terms for loose traffic are *casual* calling, *random* billing, *thrifty* billing, or *LEC-billed* traffic.

Besides the confusing and annoying arrangement of receiving two bills for long distance, the real problem with loose traffic is that it is very expensive. I have seen domestic long-distance rates as high as $7 per minute, but a typical rate is about $0.30 per minute.

Long-distance rates are based on this formula:

$$\text{gross rate} - \text{discount} = \text{net rate}$$

Loose traffic rates are high because the calls get no discount. Customers end up being charged the *gross rate,* also known as the *tariff rate.* Table 8.1 compares the high cost of loose traffic to the cost of long distance correctly billed on a long-distance bill. The figures are based on the sample phone bill in Chapter 4. Telecom consultants commonly use this format; notice the additional savings attributed to having calls billed in 6-second increments instead of full-minute increments. (See Table 7.1 for a detailed explanation of billing increments.)

Loose traffic is the single most common telecommunications inefficiency. Customers overpay their carriers large sums of money each month, and the

Table 8.1 The High Cost of Loose Traffic

CURRENT MINUTES		CPM*	CURRENT COST	NEW CPM	NEW COST
Intrastate Florida	24	$0.25	$6.00	$0.085	$2.04
Interstate	79	$0.25	$19.75	$0.095	$7.51
6-second billing incremental difference				8%	−$0.76
Total:			$25.75		$8.78
Monthly savings:					$16.97
Annual savings:					$203.62

*CPM = cost per minute

carriers do little to prevent it. In fact, their hands-off customer service practices encourage erroneous billing.

Why all the loose traffic?

Loose traffic occurs for a number of reasons. Sometimes it is the customer's fault; but usually it is the fault of one of the phone companies. Regardless of who is at fault, customers pay double or triple what they would normally pay for these long-distance calls. The problem should be corrected immediately.

New lines added by the customer

Because of the recent explosion in the use of computer modems and fax machines, customers regularly add additional phone lines to connect to these devices. When you order a new phone line from your LEC, the company always asks which long-distance carrier you want assigned to the new line. *If you fail to notify your long-distance carrier you will have loose traffic.* The line will not bill on your master long-distance account; instead, the calls on this line will bill on your local bill at high nondiscounted rates.

In the sample local bill in Figure 4.2, Acme Manufacturing needed two new phone lines to facilitate its new computerized part ordering system. When ordering the phone lines from Telephone Company A, Acme specified that both lines should have Telephone Company B as the long-distance carrier. Because it never informed Telephone Company B of the new lines, the long-distance calls on these lines are billed on the last few pages of the Telephone Company A local bill.

PIC code errors

Another reason loose traffic may appear has to do with phone company errors. The local carrier controls which long-distance company is a customer's PIC. Each long-distance carrier has its own PIC code, which is entered into the local carrier's central office and into its billing computers. (See Chapter 10 for a list of common PIC codes.)

If an overworked phone company billing representative accidentally enters the wrong PIC code for your lines, your long-distance calls will be handled by the wrong carrier. These calls will be billed on your local bill. Many carriers have multiple PIC codes, and you must ensure that the correct one is in place. Certain AT&T customers use 732 as their PIC code, but if the more common AT&T PIC code of 228 is used, the calls still might bill on the local bill.

Mismatch at the central office

Even if the PIC code is correct in your local telephone company's billing system, it may be incorrect at its central office. Since the billing computers and central office computers are usually separate systems, mismatches frequently occur. In this case, loose traffic might appear on your bill. This problem is especially prevalent when you switch long-distance carriers. Local carriers are notorious for changing the PIC in the billing system but failing to do so at the central office. Of course, the end result is that the customer pays the old rates for another month or two until somebody figures out why the change never occurred.

PIC freeze

Once a customer is satisfied that his lines have the correct PIC code, it is a good idea to request a *PIC freeze* with your local carrier. This "freezes" the PIC choice and prevents anyone from changing your long-distance carrier again unless the company has written permission from you.

Slamming

Slamming is the fraudulent practice of changing someone else's long-distance carrier without that person's permission. This is a common practice in the industry, especially among entrepreneurial start-up long-distance companies and multilevel marketing long-distance companies. If your long distance suddenly starts to appear on your local bill and you do not recognize the carrier, you have been slammed.

Slamming methods   The latest slamming techniques are becoming increasingly creative. Fraudulent carriers create sweepstakes with a free car or a cruise as the grand prize. To enroll in the sweepstakes, you fill out a small card from a countertop display found in convenience stores and restaurants. If you read the fine print on the card, you will find that you have just agreed to switch your long-distance carrier. (Do not count on taking that free cruise anytime soon.)

One of the most creative techniques has to do with the name of the long-distance carrier, as in the case of long-distance companies called "I Don't Care" and "I Don't Know." When a new customer orders lines from a local carrier, the local carrier's representative asks, "Who do you want as your long distance provider?" If the customer replies, "I don't care," then he gets his long-distance from the I Don't Care Long-Distance Company.

Another company that is successful in securing new customers fraudulently is Hold, Inc. That company's telemarketer calls you for an innocent-sounding survey, then suddenly asks "May I put you on hold?" If the person says "yes," then his long distance is switched to Hold, Inc. If a customer denies choosing Hold as her carrier, the Hold customer service representative plays back the recorded conversation to prove the customer did say "yes" when asked "May I put you on Hold?"

Slamming rights   If you have been slammed, you should not pay the charges for the first 30 days. The new FCC rules effectively give you a free month of long-distance service. According to FCC ruling 00-135, released in May 2000, you do not have to pay anyone for the first 30 days of calling. After 30 days, you must pay for the calls, but you are only responsible to pay your original carrier according to its rates. This is true even if the slamming company is still the carrier for the calls. If you have already paid the bill, the slamming company must pay your authorized carrier 150% of the charges. Your carrier is then supposed to issue a credit to your account.

Capturing loose traffic

If you have loose traffic, immediately inform your local carrier and your long-distance carrier. To get rid of loose traffic, the following steps, listed in order of urgency, should be followed:

1. *Change your PIC code.* Make your local carrier change the PIC code on your lines, both in its billing system and at the central office. Then have the company put a PIC freeze on all lines.

2. *Add the lines to your long-distance account.* Ensure that your long-distance carrier is aware of the line numbers. Make sure it adds the line numbers to your main account.

3. *Dispute the charges.* Dispute the charges with your local carrier. You have the option of withholding payment for these charges. The local carrier will not disconnect your local lines for nonpayment of another carrier's charges.

4. *Negotiate a refund.* Negotiate a refund of the overcharges with the carrier that charged you. If the loose traffic is due to a carrier error, insist that it issue an invoice credit equal to 100% of the charges.

The carrier will probably refuse to issue a full refund, but it will agree to rerate the traffic and issue a partial refund.

Although these are simple steps, many things can go wrong when trying to eliminate loose traffic. It has been my experience that a business with 10 or more locations will have loose traffic almost every month. A wise customer checks his bills every month for discrepancies, especially loose traffic errors.

Loose traffic rerate credits

If a customer has had his long distance billed as loose traffic, he is usually entitled to a refund. Unless the problem is the customer's fault, customers should not be required to pay more than they would have normally paid if the long-distance calls had been billed correctly. Loose traffic happens for a variety of reasons, and it may be impossible to figure out how it happened and who is at fault. If you cannot convince the carrier that it is the company's fault that you were overbilled, the carrier will resist giving a refund. The customer should steer the negotiation away from faultfinding and concentrate on the fact that the rates paid were too high and unfair.

Loose traffic may only involve two carriers: your local carrier and your authorized long-distance carrier. At other times, however, three carriers may be involved: the local carrier, your authorized long-distance carrier, and another long-distance carrier. If you have been slammed, however, it is likely that a fourth company has joined the party—a billing company. Billing companies such as USBI and Enhanced Services Billing (ESBI) are legitimate companies that handle the billing for fraudulent companies such as NOS and Hold. What follows is an example of how carriers operate using Luigi's Automotive Supply, a fictional Los Angeles company.

Luigi's local carrier is SBC Communications. Sprint is his long-distance carrier. A representative from Scamco Long Distance places an order with SBC to switch Luigi's long distance to Scamco. Because Scamco is a small new company with no billing agreement with SBC, Scamco has USBI process the billing. USBI represents hundreds of small telecom carriers and has a shared-billing arrangement with SBC. SBC is happy to make the change because it keeps a portion of the billing. In this fictional, but nonetheless realistic, example Luigi has four phone companies to deal with: SBC, Sprint, USBI, and Scamco. Table 8.2 shows how Luigi's long-distance cost has drastically increased as a result of the slam.

Table 8.2 Loose Traffic Cost Comparison After Being Slammed

CURRENT MINUTES		SCAMCO		SPRINT	
		CPM*	Current Cost	New CPM	New Cost
Intralata	1,564	$0.300	$469.20	$0.065	$101.66
Intrastate CA	357	$0.300	$107.10	$0.095	$33.92
Interstate	894	$0.300	$268.20	$0.095	$84.93
6-second billing incremental difference				8%	−$17.64
Monthly account fees			$5.00		$0.00
Total:			$849.50		$202.86
Monthly difference:					$646.64
Annual difference:					$7,759.62

*CPM = cost per minute

The key to negotiating a loose traffic rerate credit is to focus your energy on the company that actually billed the calls. It made money at your expense and should be willing to refund a portion of it. In the example above, Luigi should call Scamco and request a refund. He should ask that the calls be rerated back to his Sprint rates. With a *rerate credit*, Scamco refunds the difference between its rates and Luigi's actual rates with Sprint.

Table 8.3 shows the calculation of the rerate credit, based on being slammed for 6 months. Carriers rarely offer you a detailed report like this, but you can use this type of format to calculate your own refund. Most carrier representatives do not want to calculate an exact credit and prefer to offer a ballpark credit instead. Make sure the refund also includes taxes, fees, and interest. If you do secure a refund, it will appear on your local bill usually within two billing cycles. In addition to the rerate credit from the unauthorized carrier, it is sometimes possible to negotiate an additional courtesy credit from your own carrier to make up for your time and effort invested in correcting the problem.

If Scamco is uncooperative or unavailable, Luigi can try for a refund with USBI. That company will probably say "We are very sorry you can't get in touch with Scamco, but we won't give you a refund; after all, we're just the billing company, not the carrier." Luigi may be out of luck, but he can still try for a refund with Sprint.

Table 8.3 Rerate Credit

With Scamco, you paid:	$849.50
for this many months:	6
Actual amount paid:	$5,097.00
With Sprint, you would have paid:	$202.86
for this many months:	6
Corrected amount:	$1,217.16
The difference is:	$3,879.84
Taxes paid:	$436.48
Interest (1.5%):	$64.74
Total amount of your rerate credit:	$4,381.07

Besides promising "to give excellent customer service," one of the standard sales pitches of long-distance sales representatives is that "We will monitor your traffic every month." In Luigi's example, if Sprint had truly monitored his account, it would have noticed that the lines were no longer billing with Sprint. It also might have noticed that he used to bill $200 per month but now bills nothing. In the case of a larger account, Sprint would have assigned an account manager who probably would have noticed the missing call volume.

The full recourse option

If the customer cannot negotiate a refund, a full recourse of the charges can be requested with the local carrier. Explain to your local carrier that you are disputing the full amount of the charges billed by the fraudulent company. Be sure to exclude that amount from payment of your local carrier's bill.

The local carrier then notifies the fraudulent carrier that the charges are being disputed, and the fraudulent carrier has a limited time (usually 60 days) to respond. Fraudulent carriers usually do not respond, and the local carrier credits the customer's bill in the full amount.

Unethical phone companies rarely fight these disputes. In fact, many fraudulent companies are so eager to avoid customer complaints to the FCC that could result in stiff fines that they readily offer refund credits. Their phone greeting is practically "Thanks for calling Scamco, would you like a refund?"

Collect calls

Collect calls are handled by AT&T, WorldCom, Sprint, and a host of collect call niche providers. The charges for these calls usually appear in the last pages of the local bill.

Collect calls are fairly straightforward: You call collect and the person you called is charged. Encourage your employees to use 800 numbers or calling cards instead of calling collect. You can also block collect calls with the local phone company. This forces the caller to use another method to complete the call. But this is not a solution for everybody. Organizations such as law enforcement, hospitals, bail bondsmen, and lawyers regularly receive important collect calls.

900 calls

900 calls are expensive because the caller is paying for the information given by the 900 provider in addition to the long-distance charges associated with the call. Almost all 900 calling is billed on the local bill. In this way, the call is handled much the same as collect calls. 900 services are usually provided by the big long-distance companies. If you call a 900 number, you will probably see a charge from AT&T on your local bill.

900 calling has a well-earned stigma, but some of the calling is legitimate, such as technical support centers that use 900 numbers. If you have determined that your business does not need 900 calling, call your local phone company and have it block all 900 and 976 calling. 976 numbers function like 900 numbers but are based in your local market.

The block is ineffective against some 900 numbers, however, because they are accessed by dialing an 800 number first. 900 service providers are aware that most businesses block 900 dialing through their PBX or through the local carrier's central office, so they have invented a way to get past this obstacle. A caller dials an 800 number to get past the PBX, and then the call is transferred to the 900 number.

Miscellaneous monthly fees

If loose traffic, slamming, collect calls, and 900 calls are not bad enough, local bills are now fair game for a whole host of miscellaneous fees. The charges already described are all usage-based, but fees are a fixed expense each month. I have seen businesses waste thousands of dollars a year on fees that should have been canceled.

Some fees are legitimate, such as charges for voice mail, Internet access, and Web site hosting. Vendors that supply these services choose to do their billing through the local phone company because it makes collecting their money easier. As long as the customer verifies the charges on the bills each month, misbilling should be minimal.

A big problem for customers is that phone companies charge fees as a part of almost every service. Customers moving their loose traffic from their local bill to their main long-distance account will still be billed a monthly service fee of $5 to $20 just to maintain an account with the old carrier. Some carriers bill a monthly fee for each individual line on the account. Not only is it important to move the traffic, it is also necessary to inform the carrier to cancel the account.

Cramming

Cramming is the process of adding services or fees to a customer's phone bill without permission. The services are often legitimate, but the customer does not want them. A local phone company representative may have added them intentionally or accidentally. Customer service representatives are often paid a commission on each additional service they sell a customer.

Another way to accumulate fees is to make collect calls or 900 calls. Some 900 numbers, when called just one time, will enroll the caller into a monthly "membership" program. If one of your employees calls a psychic line one time, you may be enrolled as a member. Your local bill will then include a $50 membership fee each month.

Many unethical companies add monthly fees to your local bill and provide nothing in return. They deceptively give the fee a legitimate sounding name, such as "network management" or "call reporting." When the average accounts payable clerk sees the charge, she simply pays the bill rather than question the charges.

The key to avoiding or reducing the risk of cramming, slamming, collect calls, and 900 calls is to spend a few minutes each month scanning your local bills. Large companies should consider hiring an outside consulting firm for a telecom audit once a year. Discrepancies should be corrected immediately. Be firm with carriers and insist on refunds.

9

Managing your payphones

Many businesses have on-site payphones. The local phone company either owns the phones or they are customer owned. If the local phone company owns the phones, then the business owner normally pays a monthly fee to the local phone company for the payphone. The payphone bill is almost identical to a local bill for a regular business phone line. The charge for the payphone is usually about the same as the charge for one flat business line: about $40 per month.

Eliminating payphone bills

The problem with paying the local phone company to have a payphone at your site is that you are paying the company so it can earn money from your employees. Not only does the carrier earn $40 a month from you, but it gets all of the coin revenue and probably receives a commission on all operator-assisted calls, collect calls, calling card calls, and long-distance calls from the payphone.

The "$4 rule" has long been a standard with payphones. If the payphone is earning $4 or more per day in coin calls, then the local phone company will discontinue billing you the monthly fee and instead will pay a commission on the calls.

A typical business has multiple payphones at one facility. If most of the payphones are meeting the $4 minimum, then an overall commission agreement may be negotiated with the local phone company. Commissions on payphone calls vary from 5% to 25% of both the coin and long-distance revenue. The commission checks are usually paid monthly or quarterly.

If the local phone company is unwilling to waive the monthly fee for the payphone due to the low amount of calling, you may consider canceling the service altogether. If you have multiple payphones in one location that are not producing significant call volumes, consider reducing the number of phones until the $4 minimum is met.

Saving money on semipublic payphones

In many states, a site owner does not have to pay for a *semipublic* payphone. There are varying definitions for the term, but usually semipublic means the payphone is accessible to the general public. The classic example is of a payphone at the back of a loading dock. Even though the loading dock itself may close for business at 5 P.M., the payphone is still available for someone walking by. In this example, the business should be able to get the local phone company to stop charging it for the payphone each month. On the other hand, the phone company may just decide to remove the phone at that point.

Commissions directly from the operator services provider

In many markets, the site owner can choose the long-distance provider for each payphone on the premise, even if the physical payphone is owned by the local phone company. Another way to impact your bottom line with payphones is to negotiate a separate commission agreement directly with a long-distance carrier. The local phone company still provides the payphone, whether or not you pay for it each month.

In this case, the long-distance carrier is functioning as an operator services provider (OSP). The OSP handles most operator-assisted calls, and the site owner is paid a monthly commission by the OSP. The commission rate is 5% to 25% of the operator-assisted calls only; the local carrier keeps all the coin revenue.

Saving money by using a customer-owned payphone

Besides using a payphone provided by the local phone company, a business may decide to use a private payphone company, or purchase and install its own payphone. Either way, the principle is the same. Many convenience store chains use private payphone companies such as the People's Telephone Company, one of the largest private payphone companies.

The private payphone company installs and maintains its own payphone at the convenience store. If the local phone company already has a payphone on-site, which is usually the case, the private payphone company requests that it be removed. The local company will require a letter of agency signed by the site owner prior to honoring any of the private company's requests.

Once the old payphone is removed, the private payphone company orders a line from the local phone company and physically connects its payphone to the line. All installation costs should be absorbed by the private payphone company. The company should also pay the $40 bill for the line each month.

Before removing its payphone, the local phone company will probably send a sales representative out to the site owner to try to convince him not to change anything. Most of the time it is too late, because private payphone companies usually sign 5-year contracts with their customers prior to contacting the local carrier. Customers considering signing one of these contracts should contact their local carrier first. If the local carrier can offer a similar commission check each month, then the site owner should not change. Local carriers normally offer commissions on coin revenue only, not long-distance revenue.

As expected, the private payphone vendor will pay the site owner a monthly commission on both the coin calls and long-distance calls. To handle all of the operator-assisted calls and long-distance calls, the private payphone company contracts an OSP, such as AT&T, Sprint, or Opticom, one of the leading independent OSPs.

The payphone surcharge

Prior to the Telecom Act of 1996, payphone owners received no compensation for 800 calls from their payphone. Their $1,000 payphone equipment was being used for free, and the "free caller" tied up the phone, preventing a paying customer from using it. Today, this has changed. If a caller dials an 800 number from a payphone, the long-distance carrier handling the call must pay the payphone owner $0.28 for each call. This small amount of revenue slowly trickles in, but in the high-overhead payphone business, revenue from the surcharge makes all the difference.

10

Customer service records

The customer service record (CSR) is a copy of how a customer's record appears in the local carrier's computers. Like other computer records, the CSR is arcane and not much fun to look at. However, a complete telecommunications audit should include at least a cursory review of your CSRs. Most local phone bills lump multiple charges under one heading labeled "monthly service" but the bill does not itemize the charges. This chapter explains each item on a sample CSR, lists the most prevalent CSR errors, and lists the most common codes used in CSRs. The main value of being able to interpret a CSR is that you can see, in detail, exactly what charges are being billed.

Some local phone companies, such as Pacific Telesis, send their customers one copy of the CSR each year. Most carriers will provide a CSR copy in a few days at no charge or for a small fee.

Universal service order codes

The CSR is a database record that uses universal service order codes (USOC) to describe each detail of your account. USOCs were used before divestiture, when all of the RBOCs were still part of the Bell System, so many of them are still consistent today. Independent LECs such as GTE and SNET use CSRs but their USOCs differ from the RBOCs.

Your monthly telephone bill is generated based on the items in your CSR. Each item is billed according to the rate assigned to that USOC. If the USOC is incorrect, your phone bill will be inaccurate, and you will either be over- or undercharged. This is how one flat-rate business line with touch-tone service will appear on a CSR:

1FB	$20.00
TTB	$5.00
9ZR	$8.30
Total:	$33.30

1FB is the USOC for one flat-rate business line. The USOC for a measured-rate business line is 1MB. TTB is the USOC for touch-tone business. 9ZR, if itemized on the phone bill, is the FCC line charge, which is also called the end user common line charge (EUCL). The EUCL rate is raised regularly, and, ironically, this money does not go to the FCC. This fee goes straight to the local carrier and is more accurately described on some bills as the "FCC-approved line charge."

These are the most common USOCs, but thousands of others exist, and new ones are invented daily to describe carriers' new offerings. Appendix 10A contains a list of 100 of the most commonly-used USOCs, but keep in mind that USOCs are not universally used by each carrier. Most LEC customer service representatives will take time to explain the details of your CSR. If you plan to review a large number of CSRs, you should try to sweet-talk your LEC representative into giving you its internal USOC dictionary.

Armed with a list of USOCs, and a little patience, you should be able to interpret your own CSRs and audit them for accuracy. If you have complex services and want to be absolutely sure your records are accurate, hire a consultant to perform a detailed audit of your CSRs. Because CSR auditing is so tedious, the consultant will probably charge an hourly rate in addition to 50% of the monthly savings and refunds implemented on your behalf.

Sample CSR

The CSR has four different sections: header, list, bill, and service and equipment. The header section is repeated on each page of the CSR and contains information such as account number and bill date. The list section contains information about how the customer will be listed in the white pages, yellow pages, and directory assistance. The bill section contains the billing address, billed name, and tax jurisdiction. The final section, service and equipment, details all of the lines and services the carrier is providing for the customer.

Figure 10.1 is a sample CSR for a fictional customer in New York. The customer has two measured-rate business lines. Most of the RBOCs use a similar format for their CSRs. The following list offers a detailed explanation of the most relevant parts of this CSR. In the sample, many blocks are empty because there is no pertinent activity to document.

Account number   This is the main telephone number followed by the three-digit customer code. Telephone companies use different customer codes to separate the records of this customer from the records of the previous customer who had this phone number. Some phone company CSRs use the code BTN, which stands for "billed telephone number," for the main telephone number.

Class of service   This is usually 1FB or 1MB, which stand for one flat-rate business line or one measured-rate business line, respectively.

Directory   This block is used as a reference point, indicating which letter of the alphabet this business listing falls under in the directory.

Page   The page number of the CSR.

Bill period   This shows the most recent bill cycle.

Record statement   Depending on the carrier and customer's specific services, the CSR may have up to seven sections (or segments): Account, Line & Station, Key System, Special Service, Extra Listings, Account Summary, and S&E Cross Reference.

Print date   This is the actual date this paper record was printed.

Print REA   This block explains the reason for printing. BD indicates the record was printed because of the bill date.

CUSTOMER SERVICE RECORD

Telephone Company C

ACCOUNT NUMBER
212-555-1000-001

CLASS OF SERVICE	DIRECTORY	PAGE
1MB	A	1

BILL PERIOD	RECORD STATEMENT ACCOUNT	PRINT DATE	PRINT REA
OCTOBER 1 - OCTOBER 31 2001		10-07-01	BD

QUANTITY	SERVICE	DESCRIPTION	L	ACTIVITY DATE	TOTAL	T	A
	LN	ACME MANUFACTURING		05-01-97			
	LA	101 MAIN & 10013		05-01-97			
	SA	101 MAIN, MANHATTAN, NY 10013		05-01-97			
	LOC	BLDG 5		05-01-97			
	YPH						
		----BILL					
	BN	ACME CORPORATION					
	BA	250 ELM					
	PO	DALLAS, TX 75376					
	CCH	2					
	TAR	2					
		----S&E					
2	BSX	(Calling Card)	I	05-01-97		2	
1	LUD	(Local Usage Discount Plan)	I	10-08-99		1	
1	1MB	/DES RJ21X/GST / TBE B					
		/PIC TCE/PCA FN 08-09-99					
		/LPIC TCE /LPCA FN 01-20-99					
		+++FCC LINE CHARGE					
		(Monthly charge for dial tone)	T	05-01-97	16.23	1	
1	RJ21X	/DES TN 1000, 1001		05-01-97			
		(25-Line Connector Jack)	T	05-01-97		4	
1	TTB	(Touch-tone)	T	05-01-97		1	
1	ALN	/TN 555-1001					
		/DES RJ21X/GST / TBE B					
		/PIC TCE/PCA FN 08-09-99					
		/LPIC TCB /LPCA FN 01-20-99					
		+++FCC LINE CHARGE					
		LINE CHARGE++ (Additional Line)	T	05-01-97	16.23	1	
1	TTB	555-1001 (Touch-tone)	T			1	
1	MNTPB	/TN 1001 (Basic Wire Maint)	T	07-15-97	3.50		
	HTG	1 555-1000, 1001	I				
		----CALLING CARD					
	CHN	JIMMY DOE	I	05-01-97			
	CHN	DONNA DOE	I	05-01-97			
		----IN SERVICE					
3	9ZR	FCC LINE CHARGE			24.81		

RECORD STATEMENT ACCOUNT	CSG	ACCOUNT NUMBER
		212-555-1000-001

Figure 10.1 Sample CSR.

Quantity   This column shows the quantity of items listed in the next two columns: service and description.

Service   This shows the USOCs that correspond to the customer's services.

Description   This column is as close as we can get to an actual plain-English description of customer services.

L   This stands for the last service order action that was performed for the item listed in the description column. The possible codes are E, I, and T, which stand for enter, in, and to, respectively.

Activity date   This is the date of the last action for the item.

Total   This column shows the charges for each item. Where no charge appears, as in the case of HTG (or hunting), it is a free service.

T   This column uses numeric codes to show the tax status of each item. The following codes determine which taxes apply to each service:

1.   Federal, state, and local;

2.   No taxes;

3.   Federal;

4.   State and local;

5.   Federal and state;

6.   State.

A   The activity column contains an asterisk when an item has been changed since the printing of the last CSR.

LN   Listed name. This column shows exactly how the business will be listed in the white pages and directory assistance.

LA   Listed address. This column shows how the address is listed in the white pages and directory assistance.

SA   Service address. This is the physical location of the phone lines.

LOC Location. This column indicates the floor or building number where the service is located. This field helps telephone company technicians locate the physical location of the service.

YPH Yellow pages heading.

BILL This shows the start of the CSR's bill section.

BN Bill name. The name of the business as it appears on the phone bill. This may differ from the listed name.

BA Bill address. The actual address where the phone bill is sent. In this fictional example, the bill is sent to Acme Manufacturing's home office in Dallas.

PO This shows the city, state, and zip code of the bill address.

CCH This field indicates the number of calling card holders.

TAR This indicates the tax area of the customer's physical location and determines which taxes are in effect.

S&E This shows the start of the CSR's service and equipment section.

BSX This shows that the customer has two active calling cards.

LUD This is the USOC for a telephone company's Local Usage Discount Plan.

1MB One measured-rate business line. In the description column, "/PIC TCE" shows that Telephone Company E is the PIC. "/LPIC TCE" shows that Telephone Company E is the carrier for intralata calls. Further down in the CSR, notice that the second line has Telephone Company B as the LATA PIC (LPIC). The intralata calls will be carried by Telephone Company B, not Telephone Company E.

RJ21X This is the USOC for a common type of wall jack used by local carriers. The RJ21X jack serves as the demarcation point where the phone company's network connects to the customer's inside wiring. Note there is no charge for the RJ21X.

TTB Touch-tone business. With many carriers, this is not a free service, but this telephone company does not charge for touch-tone.

ALN Additional line or auxiliary line. The main number on a telephone account is often called the BTN (billed telephone number) and additional lines are called WTNs (working telephone numbers). A simple technique for CSR auditing is to verify that each ALN shows up as an exact repeat of the others. If they are not exactly the same, you may have found an incorrect PIC or LPIC, or you may have found hidden charges such as wire maintenance.

MNTPB Wire maintenance plan. The actual phone bill may not itemize this service. Note that MNTPB only appears on one of the two lines. This indicates that the customer is probably unaware of this charge, and it should be canceled.

HTG Hunting. In this example, if the calls are not answered on 555-1000, the call is forwarded to 555-1001.

CHN Card holder name. This shows the name of the employee who has been assigned a calling card. If you recognize the name of an ex-employee, cancel the card to avoid fraudulent charges. Most CSRs do not show names.

9ZR This shows the number of FCC line charges. In this example, the customer is being charged three 9ZRs but only has two lines. This error should be corrected, and Telephone Company C should issue a refund.

CSR errors

The following items are the most common errors that can easily be located when auditing customer service records. Each item is an example of local telephone company overbilling, wherein the customer should be entitled to a refund.

Wrong PIC

To keep track of which long-distance carrier a customer is using, the local carrier uses the PIC code of that carrier. The PIC code may be expressed numerically on the CSR as in "222" or with letters, as in "MCI." AT&T's most commonly used PIC code is 288. The PIC code is stored in the local carrier's central office and in its billing records. It will be printed on the CSR.

If a customer's phone lines have the wrong PIC code, his long-distance traffic will be routed to that carrier. This is a very common problem, especially in the case of slamming, when a fraudulent carrier changes a customer's PIC without the customer's authorization. When auditing CSRs, you must check the PIC code on each line. If it is wrong, call your local carrier and have the company change it. Call your long-distance carrier if you do not know its correct PIC code. Appendix 10B lists common PIC codes.

Wrong LPIC

As a result of the Telecom Act of 1996, customers can now select their carrier for intralata calling. On the CSR, the code LPIC appears and is followed by the PIC code for the carrier. Many customers have moved their intralata traffic to their long-distance carrier, so their PIC code and LPIC code should be the same. If the wrong LPIC code is on the CSR, your calls will be handled by the wrong carrier. In the sample CSR in Figure 10.1, the second phone line shows "/LPIC TCB," which means the intralata calls from that line will be carried by Telephone Company B. The customer should call Telephone Company C and have the company change it.

Too many 9ZR charges

As stated above, the 9ZR is the USOC for the end user common line charge, which is about $9 per line. Sometimes, the number of 9ZR charges is greater than the number of lines. This is especially true for large Centrex accounts that bill the 9ZR as one single-line item separate from the line charge. When auditing your CSR, count the number of lines and the number of 9ZR charges. If you are being overbilled, contact your local telephone company.

Wrong tax area

The code TAR indicates which tax area you are in. Taxes are based on where the customer is physically located. Some customers have a city address but are located outside the city limits. Such a customer should be exempt from any city taxes. In other cases, the local carrier enters the tax area of the billing address instead of the physical address of the customer. For example, a company based in downtown Chicago has a manufacturing facility in the suburbs. The taxes are lower in the suburbs, but if the phone company enters the wrong tax area, the customer will be overbilled.

Incorrect hunting sequence

HTG is the code for hunting service. If an incoming call finds your main number busy or gets no answer, hunting allows the call to automatically transfer to another line. If the second line is busy or unanswered, the incoming call hunts for another line. At the end of this "hunt group" of lines, the call will "rollover" back to the first line. Hunting is designed to prevent a business from missing out on important business calls.

A common error with hunting is that one of the numbers in the hunt group may be an old number that is no longer in use. In this case, the call will not work. Another problem is with the rollover feature. At the end of the hunt group, the call should be transferred back to the first number. If this is not set up properly, calls will be lost.

Hidden wire maintenance charges

Many local telephone bills do not offer an itemized list of all charges. Even simple phone bills with only one or two lines may contain hidden charges. You must check the CSR for these charges. The most common hidden charge is for wire maintenance. It is very common for a phone bill to simply say "monthly charge for local service" but the CSR lists MNTPB, the code for wire maintenance plans. If the customer has not ordered wire maintenance, this charge should be canceled.

Wrong mileage—first 1/4 mile

Point-to-point data circuits are billed according to the bandwidth and mileage of the circuit. The rate for the first 1/4 mile is higher than the rate for additional 1/4 miles. A 5-mile circuit will, therefore, be something like this:

First 1/4 mile:	$30
19 additional 1/4 miles @ $20:	$380
Total:	$410

If the data entry clerk makes an error when provisioning the above circuit, the customer may be billed each 1/4 mile at the higher "first mile" rate.

Wrong mileage—too much mileage

Another common error has to do with the exact mileage, which should be calculated according to the "airline mile" distance between the two local serving offices (LSO) at the ends of the circuit. If the carrier calculates the mileage according to the physical address of the two sites instead of the mileage between the two LSOs, you will be overbilled.

If you do not have access to telecom pricing software, you can double-check the mileage by giving the NPA-NXX (area code + prefix) for each location and requesting a new detailed circuit price quote from the carrier. Many carriers share software, so if you feel your current carrier may not be truthful, you can get the same information from another carrier.

Wrong mileage—double billing

Data circuits crossing LATA boundaries are usually provided by an LEC and an IXC, or long-distance carrier. The bill is handled by one of the carriers, normally the IXC. Sometimes, both carriers provide a bill for their percentage of the circuit. The customer might be billed 60% of the circuit by U S West and 40% of the circuit by WorldCom. The ratio is determined according to mileage. If 60% of the mileage is provided by U S West, then the customer's rate will be multiplied by 60%. Unfortunately, some customers end up being billed 100% by each carrier. To ensure that you are not being overbilled, match each CSR and each phone bill to your corporate network diagram.

Hanging circuit

Each data circuit must connect two points. If a customer disconnects an unneeded circuit, she should no longer receive a bill for the circuit. But sometimes, the carrier only disconnects one of the two locations. It is difficult to catch this error by looking at the phone bill alone. This error is very obvious on the CSR, however. If the CSR reads CKL 1 (circuit location 1) and there is no CKL 2, you have found a hanging circuit. The LEC should completely disconnect the circuit, stop the monthly billing, and issue a refund.

You should also check the addresses at each end of the circuit. A travel agent had dedicated lines to an airline. The agent stopped selling tickets for that airline but never canceled the billing for the dedicated lines. When the CSRs were audited, the airline's address showed up as CKL 2. The customer knew he was no longer doing business with the airline, so he canceled the circuits, saving the business about $3,000 per month.

Term plan error

Signing a 12-month term plan agreement will discount voice or data service pricing by 5% to 15%. Longer-term plans will generate greater discounts. Carriers frequently enter the wrong term plan on a CSR, resulting in missing discounts for the customer. To verify discount amounts, check the original term contract with the actual CSR. If the given discount is lower than the contracted discount, the LEC should issue a refund and correct the problem going forward. If neither customer nor carrier can produce a copy of the original contract, you may be out of luck. However, some have used this situation to eliminate an existing term commitment with the carrier.

Sliding scale line rates

Local accounts with more than 12 lines may qualify for sliding scale pricing. This is especially true with Centrex pricing. The billing may work like this:

First 25 Centrex lines @ $20 each:	$500
Next 100 Centrex lines @ $15 each:	$1,500
Next 100 Centrex lines @ $12 each:	$1,200
Total:	$3,200

Auditing the CSR might reveal that all 225 lines are being billed at the higher $20 per month rate. In this example, the customer would pay $4,500 per month instead of $3,200. This customer is entitled to a significant refund. To detect this error, the auditor must be familiar with the original contract terms or be willing to wade through the actual tariff to determine how the lines should be priced.

Loose calling cards

Few customers use the calling cards provided by their local telephone company. Lower rates are available through long-distance carriers. When a business changes calling card providers, it sometimes fails to cancel the old cards. A review of the CSR may reveal that active calling cards are still on the loose. This will not be evident by looking at the phone bill, unless someone uses the cards to make calls. Old cards should be deactivated anyway to avoid the risk of future billings if the cards are used by ex-employees or someone else.

Unused voice lines and data circuits

A very valuable piece of information on the CSR is the service address. Companies with multiple locations will often find, after they audit their CSRs, that they are still paying for lines at closed locations or at an ex-employee's home. Some businesses detect this problem many years after they quit using the lines. They may have even placed a disconnection order with the phone company. If you have documentation to prove that the lines were canceled, you are entitled to a refund. Most carriers and customers fail to keep good records, however, and the customer will never get a refund.

Chapter summary

Auditing customer service records is tedious and often unfruitful. The errors are not easy to locate, and if you do locate an error, it is often difficult to get your carrier to understand the error, correct it, and refund the overcharges. Most CSR errors are minuscule and only amount to a few dollars a month. However, if the problem has occurred on multiple lines and dates back many years, you will be entitled to a very large refund. A CSR review is a necessary component of a total telecom audit, but if your project is overwhelming, hire a consultant to do the work.

To learn more about CSR auditing, I recommend the book *Telecommunications Expense Management* by Michael Brosnan, John Messina, and Ellen Block. It is probably the best book available on CSR auditing.

Appendix 10A lists the most commonly used USOCs. Appendix 10B lists the most commonly used PIC codes.

Appendix 10A
USOC

The following list contains USOCs, field identifiers (FIDs), and other abbreviations that frequently appear on customer service records. This list should help you understand your CSRs, but if you require a complete list, contact your local telephone company.

1FB	flat-rate business line
1FL	flat-rate business line
1FR	flat-rate residence line
1MB	measured-rate business line
3LN1S	mileage charge
9LA	FCC charge for additional line
9PZ	FCC charge
9ZC	FCC charge, Centrex line
9ZR	FCC charge
ALN	additional line
BA	billing address
BCR	call return blocking
BN1	billed name
BSX	calling card
BSXUP	calling card
BTN	billed telephone number
BTN	billed telephone number
CCH	calling card holders
CKL1	circuit location 1
CKL2	circuit location 2
CLT	additional white pages listing
CTG	circuit termination charge

D1F2X	off-premise extension
EABDF	automatic callback
ESC	three-way calling
ESM	call forwarding
ESX	call waiting
EXF	off-premise extension
FX	foreign exchange
HTG	hunting/rollover service
LA	listed address
LN	listed name
LOC	location
LPIC	intralata toll carrier
LPIC	intralata toll carrier
LSO	local serving office
MJB	special business account
MNTPB	wire maintenance
NDT	PBX service
NDZ	sliding scale charge
NFB	traffic record
NPU	nonpublished number
NW1	network interface
NYZAA	Centrex contract billing
OPX	off-premise extension
OUV	local usage discount plan
PIC	primary interexchange carrier
PIC	primary interexchange carrier
PMWV4	voice grade circuit—basic 4 wire

PO	post office
RCF	remote call forwarding
RJ11C	miniature jack
RJ14C	2-line modular baseboard jack
RJ21X	25-line connector jack
RX2	Centrex line
RXR	Centrex line
SA	service address
SDS	ISDN basic exchange service
SEQ1X	wire maintenance plan
TAR	tax area
TB2	direct incoming trunk
TCG	combination in-out trunk
TCM	outward trunk
TDN	touch-tone
TJB	touch-tone on trunks
TN	telephone number
TTB	touch-tone business
TTR	touch-tone residence
TW6	direct incoming trunk
TXG	combined in-out trunk
TXM	additional out-only trunk
VF3	foreign exchange service
VMN2X	voice mail
VOP3X	large volume discount
VOP5X	large volume discount
WMR	deregulated access line

WTN	working telephone number
WTN	working telephone number
YPH	yellow pages heading

Appendix 10B
150 Most commonly used PIC codes

The following list contains 150 of the most commonly used PIC codes. Some of the codes are outdated but will still appear on customer service records if a customer has not changed his long-distance carrier for many years. LDDS appears numerous times because it acquired many companies and inherited their PIC codes.

10001	LDDS
10002	LDDS
10004	ATX Telecommunications Services
10008	LDDS
10022	MCI
10025	Inter Continental Telephone
10031	LDDS
10033	Sprint
10055	Wiltel
10058	Schneider Communications
10066	Allnet Communications Services [Lexitel/Ldx]
10070	U.S. Long Distance
10074	US Tel
10080	LDDS Metromedia
10084	LDDS Metromedia
10087	Telecom*USA [MCI]
10088	Wiltel Communications
10088	SBS/MCI
10092	MCI
10124	Horry Telephone Long Distance
10128	Amnet

10133	MCI
10135	Network One
10142	LDDS dba First Phone of New England
10177	Sprint
10183	Telecom*USA [MCI]
10187	BellSouth Telecommunications
10188	SBS/MCI
10200	US WATS
10220	Telecom*USA [MCI]
10222	MCI
10223	Cable & Wireless
10224	Telecom*USA [MCI]
10226	One Call
10227	People's Telephone Company
10232	Schneider Communications
10236	Litel/LCI International/Qwest
10237	LDDS
10240	Deltacom
10246	Shared Communications Services
10252	Long Distance USA [Sprint]
10253	Litel/LCI International/Qwest
10256	Applied Signal Corporation
10286	Cincinnati Bell Long Distance
10288	AT&T
10290	LDDS
10291	Call Savers
10305	Motorola

10311	Metromedia Communications Corp. [SNC]
10314	McCaw Cellular Communications
10321	Telecom*USA [MCI]
10323	BC Tel
10333	US Sprint
10334	Pagenet
10348	Sprint Canada
10353	LDDS
10363	Bell Canada
10372	Pacific Bell
10375	AT&T Easylink Services
10376	Northern Telecom
10378	Tel Serv
10387	AT&T
10402	Integretel, Inc.
10407	Nuestra Telefonica
10411	American Express Travel Related Services
10421	LDDS
10428	LDDS
10432	Litel/LCI International/Qwest
10449	Merrill Lynch
10450	LDDS
10451	Litel/LCI International/Qwest
10453	LDDS
10454	Hewlett Packard
10473	LCI International dba Afford-A-Call
10488	Metromedia Communications Corp.

10492	Coastal Long Distance Services
10500	Schneider Communications
10523	LDDS
10531	Farmers Long Distance
10535	LDDS
10537	LCI International dba Long Distance Service
10538	LDDS
10543	Digitran Corp
10555	WilTel Communications
10560	TCI Communications
10562	LCI International/Qwest
10595	Vartec Telecom dba 10+ Teleservices
10614	Cincinnati Bell Long Distance
10623	National Teleservice
10624	American Telesource International
10625	EDS
10628	Star Telephone Long Distance
10650	EDS
10665	LCI International dba Teledial America
10695	Opticall
10732	AT&T
10733	LDDS/Metromedia Automated Operator
10737	LDDS/Metromedia Automated Operator
10738	LDDS Communications
10741	LDDS
10742	LDDS
10752	Excel

10754	Advantis
10755	Advantis
10756	Zeroplus Dialing Inc. (ULD)
10763	Southern New England Telephone (SNET)
10769	Omni
10771	Telecom*USA [MCI]
10772	Advantis
10777	Sprint
10780	Matrix Telecom
10786	LDDS
10789	LDDS
10800	LDDS Metromedia
10801	LDDS
10802	Electric Lightwave
10811	Intellical Operator Services (VarTech Payphones)
10812	Network Access
10818	Vartec Telecom dba 10+ Teleservices
10820	GE Exchange
10824	LDDS
10826	Telecom*USA [MCI]
10832	Telecom*USA [MCI]
10835	Telecom*USA [MCI]
10839	Cable & Wireless
10846	LDDS dba TMC of Southwest Florida
10850	Digitel
10852	Telecom*USA [MCI]
10857	Telecom*USA [MCI]

10862	LDDS
10863	Unitel
10872	Sprint
10873	Telstar Communications
10876	Telecom*USA [MCI]
10880	One Call Communications dba Opticom
10888	MCI
10898	SBS/MCI
10899	Telephone Express
10900	MCI
10909	Chadwick Telephone
10910	LDDS
10911	LDDS
10921	LDDS
10951	LDDS
10957	Uni Dial
10972	Digital Network
10980	Midwest Telecom
10986	MCI
10987	LDDS Metromedia Automated Collect Call Service
10988	AT&T Easylink Services

11

Long-distance overview and sample bill

When the telephone was originally invented and deployed, most of the phone lines only connected within a single urban area. A caller could only call someone who lived in the same city. If long-distance communication was needed, one had to send a telegraph or simply write a letter. At the end of the nineteenth century, all of this changed and the world became a smaller place. In 1881, the first long-distance call was made from Boston, Massachusetts, to Providence, Rhode Island. AT&T built a nationwide network and soon a person could call anywhere in the country.

The government viewed AT&T's network as an asset to the nation and kept a hands-off approach while AT&T built it out. For almost 100 years, AT&T operated as a monopoly in more than 90% of telephone service markets. However, a series of legal battles culminated in the divestiture of AT&T in 1984, opening the door for hundreds of other long-distance carriers. During the 1980s and 1990s, AT&T's revenues grew, but its share of the long-distance market continued to drop. While it is still the industry leader today, AT&T's share of the long-distance market is less than 50% for the first time.

What is a long-distance call?

The divestiture of AT&T clearly defined long-distance service. The Bell companies could now handle all calls originating and terminating within the same LATA. This includes both local calls and intralata calls, also known as local toll calls. AT&T would handle all calls between LATAs. As these are interlata calls, the long-distance carriers became known as interexchange carriers. In fact, one of the leading long-distance wholesalers is named IXC Corporation.

Figure 11.1 shows the LATAs in Arizona. A call within Phoenix is a local call. A call from Phoenix to another city within the same LATA, such as Kingman, is an intralata call. Both calls are normally carried by the local telephone company. A call across LATA boundaries, such as a call from Phoenix to Tucson, is a long-distance call and must be carried by the long-distance carrier.

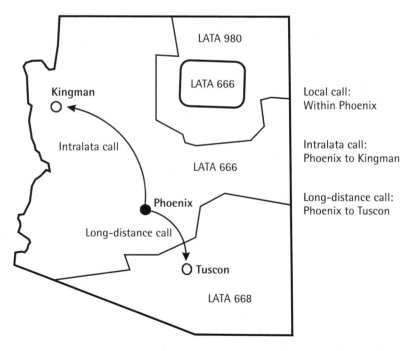

Figure 11.1 Local access and transport areas in Arizona.

How does a long-distance call work?

Standard phone lines connect an end user to the local phone company's central office. The central office computer, called a switch, interprets the numbers dialed and then routes the call to its next destination (see Figure 11.2).

When the central office detects that "1 + area code" has been dialed and determines the call is to another LATA, it knows the call must be handed off to the long-distance carrier. The central office computer then queries a database to find out which long-distance carrier has been selected for the line. The PIC code is used by the local telephone company's computers to keep track of a customer's chosen long-distance carrier. Appendix 10B lists the most commonly used PIC codes.

Once the central office computer determines that the call will terminate outside the LATA, it connects the call to the caller's chosen long-distance carrier. Although the switching equipment at the local carrier's central office may be exactly the same as the long-distance carrier's central office, the long-distance carrier's central office is almost always called a *point of presence*. That term refers to the fact that the local carrier has a complete phone network within the LATA, while long-distance carriers only have limited network points within the LATA.

When a caller needs to make a long-distance call, the local exchange carrier's central office directs the call to the nearest point of presence so the call can then be carried by the long-distance carrier. At the terminating end

| 1. Doug dials 1-520-555-5555 to reach Brenda in Tucson. | 2. The central office recognizes this as a long-distance call. | 3. The central office queries its database: "Who is Doug's PIC?" | 4. The call is routed to the long-distance carrier. | 5. The long-distance carrier transports the call to Tucson. |

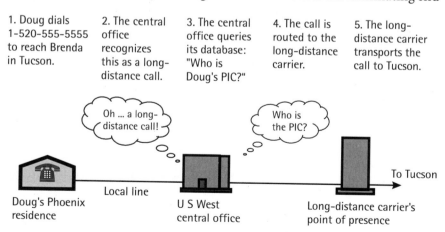

Figure 11.2 How a long-distance call works.

of the call, the long-distance carrier connects to the local carrier's central office that serves the person being called.

These multiple interconnections and handoffs between central offices take place within milliseconds. The whole process is seamless to the end user, except for the occasional faint clicking sound that may be the result of antiquated equipment in an older central office.

Long distance is now a commodity

Companies that built their own telecommunications networks such as Sprint and MCI began to win market share from AT&T. Hundreds of long-distance resellers also entered the foray. Long Distance Discount Savers was one such company. Later known as LDDS and then as WorldCom, this tiny Mississippi company soon became one of the most dominant telecommunications providers in the world in less than a decade. The 100-year dominant reign of AT&T was definitely over.

Residential and business customers now have more choices than ever for their long-distance service. Over the years, customers have left AT&T for a variety of reasons. Some prefer the more personal service smaller carriers offer. Some left because they think calls may be clearer on another carrier's network. However, most left AT&T because they have found lower prices elsewhere.

Most industry insiders agree with the statement "long distance is now a commodity." What they are really saying is that the service the end user receives does not vary from carrier to carrier. The sound quality of a long-distance phone call is so clear now that end users rarely experience any problems when making a long-distance call.

The core issues that separate one long-distance provider from another in today's marketplace are advanced features, customer service, pricing, billing, and technical support. These are the issues that heavily influence telecom managers and corporate controllers today when negotiating new service with a long-distance provider.

Businesspeople in charge of managing their telecommunications services are under pressure from within their organization to ensure three things:

1. Secure customized long-distance service that meets the specific needs of the organization.

2. Choose a carrier that will not allow any service outages.

3. Secure the lowest pricing available in the industry.

The rest of this chapter will address these concerns. By following the tips and management strategies given in the following chapters, you should be able to manage your long-distance carrier(s) in such a way that all three objectives are met.

Long-distance service

Long-distance service consists of three major service types: outbound, inbound, and calling cards. Outbound long distance is what most of us know as direct dial long distance. To complete the call, the caller dials

$$1 + \text{area code} + \text{number}$$

Inbound long distance, also known as toll-free service, refers to a caller dialing an 800 number to reach a business. It is a toll-free call for the caller; the toll shows up on the business' long-distance bill. Since the finite number of 800 numbers is running out, new numbers such as 888 and 877 are now used for toll-free calling.

Calling cards are most frequently used by callers who are traveling. Rather than use coins in a payphone or cause a charge on a host's long-distance bill, calling cards offer a caller convenience and itemized billing each month. The complex world of long-distance service is not as baffling if you understand these three categories of any long-distance calling.

How a calling card call works

Calling card long-distance rates are higher than outbound and inbound rates. The cost of a calling card call normally includes a surcharge in addition to the per-minute rate. The surcharge may be as little as $0.15 per call or as much as $2.50 per call for some older cards still in circulation. The surcharge is designed to cover the cost of setting up the call.

During the past few years, the trend is that surcharges are lower or not charged at all. A business today must choose between cards with low rates that have a surcharge, or flat-rate cards that have no surcharge. In general, a business that makes very brief calling card calls should use a flat-rate card. Businesses that make long calls are probably better off with a traditional card that charges a surcharge. On a typical AT&T pricing plan, the surcharge of $0.35 and the cost per minute is the same as the direct dial outbound rate. Table 11.1 shows how the billing works for a typical caller.

Use a discount carrier

The simplest way to cut your costs associated with calling cards is to switch to a discount carrier that specializes in the service. These niche carriers, such as VoiceNet, offer calling card rates that are usually lower than full-service carriers' rates. VoiceNet advertises heavily in in-flight magazines and uses a wide network of independent sales agents. VoiceNet's current program is a flat rate $0.149 card with no surcharge, which is one of the lowest rates in the industry.

Occasionally, other discount calling card providers spring up with rates that sound too good to be true. Be careful about doing business with these companies because their actual billed rates may be higher than their actual rates. When choosing a discount carrier, choose a stable carrier that has been in business for more than 2 years.

Once you have chosen the discount long-distance carrier for your calling cards, you should compare the cost. Normally, you can give your

Table 11.1   Calling Cards: Saving Money with Flat-Rate

	NORMAL CALLING CARD			FLAT-RATE CARD	
Call Length	Surcharge	Rate	Call Cost	Rate	Call Cost
30 seconds	$0.35	$0.10	$0.40	$0.149	$0.07
54 seconds	$0.35	$0.10	$0.44	$0.149	$0.13
60 seconds	$0.35	$0.10	$0.45	$0.149	$0.15
90 seconds	$0.35	$0.10	$0.50	$0.149	$0.22
2 minutes	$0.35	$0.10	$0.55	$0.149	$0.30
2 min., 30 sec.	$0.35	$0.10	$0.60	$0.149	$0.37
3 minutes	$0.35	$0.10	$0.65	$0.149	$0.45
5 minutes	$0.35	$0.10	$0.85	$0.149	$0.75
10 minutes	$0.35	$0.10	$1.35	$0.149	$1.49
			$5.79		$3.93
				Difference:	$1.86
				Difference:	32%

current calling card bill to the sales representative who will analyze it and offer a cost comparison. It is a good idea to then do your own comparison.

Prepaid cards

Prepaid calling cards are very lucrative for carriers, which is why they can afford to give them away as gifts so often. These cards are so lucrative because carriers get the revenue from the customer before they actually provide the service. In many cases, the carrier never does provide the service. Calling cards expire, and many are thrown away when they only have a couple of minutes remaining.

Most businesses should stay away from prepaid cards because they hurt cash flows and are difficult to manage. The expense of paying for calling cards before you use them shows up in the company's books at least 2 months before it would have shown up with traditional pay-after-you-use-them cards. Keeping track of employees who use calling cards is another drawback of prepaid cards. Once the cards are issued, you never know how much they are used because the carrier will not send you a bill that shows the usage.

If you are willing to keep track of all the users yourself, then prepaid cards may be for you. For businesses that use temporary employees or fear their employees will fraudulently use calling cards, then prepaid cards may be the best option.

A few carriers offer a rechargeable prepaid card. ATX, a regional long-distance carrier in the Pennsylvania area, offers a superior rechargeable prepaid card. With rechargeable cards from ATX, a manager can issue cards to traveling employees with a limit, such as $50 per month. If the employee tries to make more calls, he has to call the home office and ask the manager to recharge the card. Rechargeable prepaid cards are very successful in limiting employee abuse and fraud.

Once a perpetrator obtains your calling card number and PIN, he can rack up thousands of dollars in fraudulent billing in just a few days. Using rechargeable prepaid calling cards that have a limit will minimize your risk of being defrauded.

Save money by using an 800 number instead of cards

Most telecommunications cost management measures do not require any advanced knowledge of the services or billing. By taking time to review the phone bills each month and apply a little common sense, most people should be able to successfully manage their telecommunications services.

Inbound long distance

Inbound long distance has its roots in AT&T's WATS, which was more of a bulk pricing service than a sophisticated telecommunications service. Nonetheless, In-WATS service could be used by a business to allow its traveling employees and remote customers to call in for free. The business being called that signed up for In-WATS service paid for the calls. This became a significant competitive advantage for sales organizations that relied on their sales to come from inbound phone calls. Consumers are far more likely to call a business with toll-free service than if they have to pay for the call themselves.

Because inbound long distance requires more of the carrier's network resources, inbound long-distance rates are slightly higher than outbound long-distance rates. In general, inbound long-distance rates are one cent higher than outbound long-distance rates. As previously mentioned, most carriers charge a monthly recurring fee of $10 to $20 for each toll-free number. The higher rates and fees for inbound long distance ensure that the carrier's additional costs for this service are covered.

The "ring to" number

When signing up for inbound long distance today, customers must tell the long-distance provider to which number they want the 800 number connected. 800 numbers are virtual numbers. They do not have physical wires assigned specifically to each 800 number. Instead, the calls come in across a regular phone line that is specified by the customer. This number is often called the "ring to" or "pointed to" number.

A small business with five local lines would probably choose to have its 800 number pointed to its main phone number. With a simple phone system, the person who answers the phone may not know if the caller is using the 800 number or not.

The RESPORG

When the 800 number is dialed from a remote site, the public-switched network must figure out how to accurately route the call to the correct destination. Figure 11.3 shows how telephone companies route 800 calls.

The following example illustrates how the routing of 800 calls works. Mary just moved to San Diego, California, and wants to get a quote for auto insurance. She calls an 800 number she saw in a television advertisement for the Low Cost Insurance Company in Chicago, Illinois. When she dials

1. Mary dials 1-800-555-5555 from a payphone in San Diego.

2. Pacific Bell queries its database: "Who is the RESPORG for this 800 number?"

3. The call is routed to Qwest.

4. Qwest queries its database: "What is the ring-to number?"

5. The call is transferred to the Chicago office.

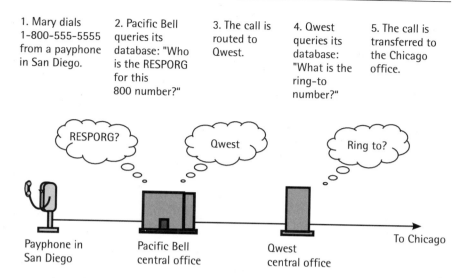

Figure 11.3 How an inbound long-distance call works.

the 800 number from her house, her Pacific Bell local central office has to determine how to route the call. The central office queries a database to find out which long-distance company is the responsible organization, or RESPORG, for this 800 number. The database responds that Qwest is the responsible organization. The call is then connected to Qwest's nearest point of presence.

Next, Qwest's central office computer must determine how to route the call. The computer queries a database and asks what phone number the 800 number is pointed to. Once the database responds with the "pointed to" number, the call is routed to the Qwest point of presence nearest to that exchange. At that point, the call is connected to the Ameritech central office, which completes the call. The call is then connected across Low Cost Insurance Company's main telephone line.

Change carriers for inbound long distance

Prior to 1993, if customers wanted to switch long-distance carriers for their 800 service, they would also have to change to a new 800 number. AT&T controlled most of the long-distance market at that time, and its rates were usually 5% to 50% higher than competitors' rates. If customers wanted to change carriers, they would have to be willing to put up with the hassles involved with changing 800 numbers. If the number was used only by company employees, the change might not have been too cumbersome. On the other hand, if the number was highly publicized and advertised, the potential lost business could far outweigh the cost savings associated with switching to another carrier.

In 1993, 800 number portability was implemented. As a result, customers can switch their inbound long-distance service to another provider but retain the same 800 number. However, a change in the RESPORG must take place. To change carriers, the new carrier requires customers to sign a change of RESPORG form, which is then sent to the old carrier and serves as a request for the old carrier to release the 800 number.

All of the carriers cooperate nationally to keep track of who is responsible for each 800 number. Even local carriers participate, because they may be the RESPORG for a customer's 800 number that is used only to carry intralata traffic. The carriers usually explain that they must charge a fee for each toll-free number a customer has so they can fund the national toll-free directory and database. The fees per toll-free number may be as low as $1 per month with WorldCom or as high as $50 per 800 number with AT&T's MegaCom billing.

Save money on 800 fees

The simplest strategy here is to switch your service to a carrier that does not bill fees per 800 number, or get your current carrier to waive the fees. Like the banking industry, the telecommunications industry earns a significant amount of fee income. If the customer has the right amount of leverage, the average long-distance carrier will waive the toll-free number fees. A customer who has just completed a term agreement with a carrier and is renegotiating a new contract probably has enough leverage to get these costly fees waived.

When negotiating contracts with carriers, it is of the utmost importance that the cost of fees be clarified before executing the agreement. During negotiations, long-distance carriers will usually steer the conversation to discuss discounts and rates. Too many customers have allowed the negotiations to end here. When they get their bill, they may be surprised to see miscellaneous fees increase the bill by as much as 30%.

As could be expected, carriers are never happy about losing business. Most carriers will fight to retain a customer once they receive a change of RESPORG form from a competing carrier. They are not allowed to refuse to give up the number, but they do not have to forfeit the business without a fight. The first tactic is to call the customer directly and try to retain the business. The incumbent carrier may offer lower pricing or other premiums such as a free month of service in an attempt to save the account. If the customer is switching due to poor service, the carrier will probably attempt to resolve the problem.

Beware of the name mismatch game   If none of the tactics mentioned works, the incumbent carrier will play the name mismatch game. Upon receipt of the change of RESPORG form from the new carrier, the old carrier will double-check the exact spelling of the customer's name. If the names mismatch only slightly, the carrier will refuse to release the number. I have seen numerous cases in which the current carrier had misspelled the company name years ago, and now that the company wants to change carriers, the current carrier will not release the number because of a mismatch.

For example, a company called Dave's Trucking and Transportation uses Sprint for its 800 service. Dave wants to switch to AT&T, so he fills out the proper forms with the AT&T representative. Sprint refuses to release the 800 numbers because it knows the account as "Dave's Trucking and Transportation, Inc."

In an era where mergers and acquisitions cause business names to change frequently, the name mismatch game can be a very effective way for a long-distance carrier to earn an additional 2 months' worth of billing.

Save money (and hassles) when switching inbound long-distance carriers
When you fill out the RESPORG forms with your new carrier, give the company a copy of your current long-distance bill. This way, the carrier can ensure the name on the forms will match exactly. Better yet, have the company send the bill copy to the new carrier along with the RESPORG form. This may save you up to 2 or 3 months of extended billing with the old carrier, as well as the frustration associated with micromanaging your long-distance carriers.

The carriers

Long-distance carriers are not all alike. Some are small, specialized carriers, while others such as WorldCom are giant carriers offering a wide variety of services. The following sections explain the types of carrier that typically offer long-distance service.

Supercarriers

AT&T, Sprint, WorldCom, and Qwest all brag that they can be a "one-stop shop" and provide all the telecom services a business customer needs. Some people in the industry call these supercarriers, because they can offer not only long-distance service but also local service, data service, and wireless service. The main advantages of supercarriers are that customers can receive one consolidated bill, and the carrier might offer some services at cost. For example, Qwest may give rock-bottom pager pricing in hopes of also securing a customer's data and long-distance business.

Regional carriers

In certain regions, a single independent long-distance carrier may earn a large share of the market. Through referral marketing, these long-distance carriers build a loyal following in their region. Later, they hope to break into the national scene. Two regional carriers that have successfully competed in the national marketplace are Frontier and McLeod USA. Regional carriers usually have low rates and are very attentive to their customers, but they struggle to manage large accounts with multiple locations.

Resellers

Supercarriers are eager to win as much market share as possible, so they allow their services to be resold by other companies. With a reseller such as Network Plus, a customer uses AT&T's network but receives a phone bill from Network Plus. During the 1990s, resellers' rates were very low so many customers switched to resellers. However, resellers are limited in the range of services they can offer, and they are not very good at troubleshooting customer problems. This is because they do not control the physical network or the actual billing records, only the *rebilling* records.

Agents

Agents are like resellers in that they do not have their own physical network of lines. Unlike resellers, they do not pretend to be an actual phone company. Instead, they are simply sales agents for long-distance carriers. Telecom technicians sometimes become agents for long-distance carriers. They are highly trusted by their customers and can earn extra commissions by selling long distance.

Most agents work alone or in a small firm. They represent numerous long-distance carriers and can usually offer the lowest rates in the industry. During a sales presentation, an agent may present proposals from two or three carriers. They are generally more loyal to their customers than to the carriers they represent. Agents usually give better customer service than the actual carriers, but an agent's recommended carrier may not always be the best carrier for the customer. Agents are likely to recommend the carrier that pays the greatest commission, which may or may not be the best carrier for the customer.

Consultants

Most agents say they are consultants, but consultants never say they are agents. Consultants are independent of carriers and receive compensation only from their clients. Agents are paid by the telephone companies they represent.

Consultants are usually people who have worked for the carriers for many years. Because of their experience, they have become experts in their specific field. The two most prevalent types of consultants are cost management experts and technical experts. Unlike agents, consultants always make unbiased recommendations to their clients.

Local telephone companies

Many local telephone companies now offer long-distance service. The rates may or may not be competitive. Before switching your long-distance service to your local carrier, compare its pricing to another long-distance carrier's pricing. The main advantage of using your local telephone company for long distance is that it simplifies your telecom environment. The consolidated phone bill is easier to manage than separate bills, and you no longer have to put up with the finger-pointing of two carriers.

Sample long-distance bill

Figures 11.4(a–c) show a typical long-distance telephone bill. This simplified bill is for a small business in New Jersey that has two lines on the account. The first page of bill, Figure 11.4(a) is very straightforward. It shows last month's charges, this month's charges, notices from

Telephone Company D

ACME MANUFACTURING
123 MAIN ST
MOUNT HOLLY, NJ 08060

Page: 1
Billing Period Ending: 10/25/01
Invoice Date: 10/26/01
Customer Number: XXXXXXXX

Summary of Charges

Balance Forward	Account Adjustments	Telephone Company D Charges	Telephone Company D Discounts	Taxes and Regulatory Rel. Charges	Current Total	Amount Due by 11/18/01
$.00	$.00	$250.79	-$10.93	$49.07	$234.47	$234.47

Important Information from Telephone Company D:

STOP — PROTECT YOURSELF FROM BEING SWITCHED TO ANOTHER CARRIER!
Call your local company today and request that a protection be placed on your line to prevent another carrier from making a change to your account without your permission.

If you have any questions about your invoice, please call Customer Service at 1-800-555-5555, or visit us at http://www.xxx.com

Fold, then Detach and Return this Portion with Your Payment

Telephone Company D

Customer Number: XXXXXXXX

Amount Due by 11/18/01
229.12
AMOUNT ENCLOSED $ _____

ACME MANUFACTURING
123 MAIN ST
MOUNT HOLLY, NJ 08060

Telephone Company D
P O BOX 5555
AUGUSTA, GA XXXXX

☐ Please check here if your address has changed.

Thank You For Using Telephone Company D.
Make Check or Money Order Payable to Telephone Company D in U.S. Dollars.
Do Not Send Cash.

Figure 11.4(a) Sample long-distance phone bill: page 1.

the carrier, and the payment coupon. The bill's second and third pages, Figures 11.4(b–c), show the usage amounts and all the charges. This customer used a total of 714 minutes for a charge of $196.33. Taxes and related charges total $43.72.

Telephone Company D

ACME MANUFACTURING

Page: 2
Billing Period Ending: 10/25/01
Customer Number: XXXXXXXX

Account Detail

BALANCE FORWARD

Description	Date	Amount
Previous Balance		$461.25
Payment Received - Thank You	3/22/01	-$461.25
BALANCE FORWARD		$.00

CHARGES

	Calls	Minutes	Amount
In-State	4.0	290.0	$54.41
State-to-State	36.0	423.0	$140.28
International	1.0	1.0	$1.64
TOTAL CHARGES	41.0	714.0	$196.33

DISCOUNTS

DISCOUNT	610-555-5000	-$5.49
DISCOUNT		-$5.44
TOTAL Telephone Company D DISCOUNTS		-$10.93
CURRENT MONTH SUBTOTAL		$185.40

TAXES/REG. RELATED CHGS.

STATE SALES TAX	$15.57
CARRIER UNIVERSAL SVC CHARGE	$18.35
PICC	$5.50
PROPERTY TAX ALLOCATION	$1.85
FEDERAL EXCISE TAX	$7.79
TOTAL TAXES/REG. RELATED CHGS.	$49.07

CURRENT TOTAL	$234.47

Itemization of Calls

ORIGINATING NUMBER: 609-555-5000

Nbr	Date	Time	*	Called Location		Called Nbr	Minutes	Amount
1	10/1/01	9:34 AM	D	PHILA	PA	215-XXX-XXXX	1.0	$.33
2	10/1/01	2:24 PM	D	WILMINGTON	DE	302-XXX-XXXX	13.0	$4.29
3	10/3/01	3:29 PM	D	PHILA	PA	215-XXX-XXXX	27.0	$8.91
4	10/4/01	9:34 AM	D	KINGPRUSSI	PA	610-XXX-XXXX	69.0	$22.77
5	10/4/01	11:53 AM	D	PHILA	PA	215-XXX-XXXX	22.0	$7.26
6	10/4/01	1:34 PM	D	PHILA	PA	215-XXX-XXXX	1.0	$.33

Figure 11.4(b)   Sample long-distance phone bill: page 2.

Calculate the cost per minute

To know exactly how much your carrier is charging you for long-distance calls, you must calculate the exact cost per minute. To do this, divide the

Telephone Company D

ACME MANUFACTURING

Page: 3
Billing Period Ending: 10/25/01
Customer Number: XXXXXXXX

Itemization of Calls

ORIGINATING NUMBER: 609-555-5000

Nbr	Date	Time	*	Called Location		Called Nbr	Minutes	Amount
7	10/6/01	8:35 AM	D	SUMMIT	NJ	908-XXX-XXXX	25.0	$4.75
8	10/6/01	2:24 PM	D	SUMMIT	NJ	908-XXX-XXXX	32.0	$6.08
9	10/7/01	9:25 AM	D	PHILA	PA	215-XXX-XXXX	27.0	$8.91
10	10/7/01	11:39 AM	D	KINGPRUSSI	PA	610-XXX-XXXX	69.0	$22.77
11	10/10/01	11:53 AM	D	SOMERVILLE	NJ	908-XXX-XXXX	225.0	$42.75
12	10/12/01	1:34 PM	D	PHILA	PA	215-XXX-XXXX	1.0	$.33
13	10/12/01	2:41 PM	D	PHILA	PA	215-XXX-XXXX	2.0	$.66
14	10/20/01	3:34 PM	D	PHILA	PA	215-XXX-XXXX	3.0	$.99
15	10/21/01	6:58 AM	D	NEWSMITHVL	PA	610-XXX-XXXX	4.0	$1.32
16	10/21/01	7:32 AM	D	ISANTI	MN	763-XXX-XXXX	5.0	$1.65
17	10/21/01	12:54 PM	D	ANOKA	MN	763-XXX-XXXX	6.0	$1.98
18	10/22/01	8:46 AM	D	UNITEDKGDM	UK	215-XXX-XXXX	1.0	$1.64
19	10/25/01	9:13 AM	D	OCEAN CITY	NJ	215-XXX-XXXX	8.0	$1.52
20	10/25/01	9:34 PM	D	PHILA	PA	215-XXX-XXXX	9.0	$2.97
TOTAL FOR 609-555-5000							417.0	$98.32

ORIGINATING NUMBER: 609-555-5000

Nbr	Date	Time	*	Called Location		Called Nbr	Minutes	Amount
1	10/1/01	8:45 AM	D	SUMMIT	PA	215-XXX-XXXX	3.0	$.99
2	10/6/01	10:34 AM	D	SUMMIT	PA	215-XXX-XXXX	21.0	$6.93
3	10/7/01	2:24 PM	D	PHILA	PA	215-XXX-XXXX	27.0	$8.91
4	10/7/01	11:39 AM	D	KINGPRUSSI	PA	610-XXX-XXXX	22.0	$7.26
5	10/10/01	11:53 AM	D	KINGPRUSSI	PA	610-XXX-XXXX	54.0	$17.82
6	10/11/01	9:25 AM	D	ANOKA	MN	763-XXX-XXXX	1.0	$.33
7	10/12/01	1:34 PM	D	PHILA	PA	215-XXX-XXXX	2.0	$.66
8	10/20/01	2:43 PM	D	PHILA	PA	215-XXX-XXXX	3.0	$.99
9	10/24/01	3:34 PM	D	PHILA	PA	215-XXX-XXXX	22.0	$7.26
10	10/25/01	9:34 PM	D	PHILA	PA	215-XXX-XXXX	9.0	$2.97
TOTAL FOR 609-555-5000							164.0	$54.12
TOTAL ITEMIZATION OF CALLS							714.0	$196.33

For a description of rate periods, please see terms and conditions.

Figure 11.4(c) Sample long-distance phone bill: page 3.

net charge by the minutes. To determine the net charge, first subtract the discount from the *gross charge*. As shown in Figure 11.4(b), two discounts are on the account that amount to 5.5% off the gross total. The bill does not clearly indicate how the discounts apply to each call type, so we must assume they apply equally. Therefore, the *net charges* for domestic calls are calculated as follows (international calls are ignored because the volume is so low):

In-state $54.41 − 5.5% discount = $51.42

State-to-state $140.28 − 5.5% discount = $132.56

After determining the true *net* cost for each call type, divide this amount by the number of minutes. The domestic call rates are calculated:

In-state $51.42/290 minutes = $0.177 per minute

State-to-state $132.56/423 minutes = $0.313 per minute

Even though the customer is receiving two discounts, the rates are three times higher than they could be in today's market. This particular bill is from a small office of a large corporation. The corporation is large and does not have a system in place to audit all of the phone bills for each location. The corporation probably has a national account with a long-distance carrier. The bill could be reduced by about $150 per month by moving the traffic to the corporate long-distance account. Employees at the office in New Jersey are either unaware of the national account or unconcerned about managing these costs.

Taxes and related charges

The next section of the bill in Figure 11.4(b) shows a number of taxes and related charges. The state sales tax varies in each state. The carrier universal service charge is a fairly recent tax. The funds are earmarked to be spent improving telecom infrastructure in schools, libraries, and rural areas.

Incorrect PICCs

As explained in Chapter 10, the PICC is to be charged by the long-distance carrier, who then pays this money to the local carrier. If a line does not have a long-distance carrier presubscribed, then the LEC will bill the PICC. The rate of the PICC is based on the type of line the customer has. PICC monthly rates are shown in Table 11.2.

The long-distance carrier has no way to verify what type of lines a customer has, so it inevitably charges the higher rate. If you have Centrex service, chances are you are being overbilled by your long-distance carrier. If

Table 11.2   Sample PICC Rates

Business single line:	$0.53
Business multiple lines:	$2.75
Business multiple lines in California:	$1.84
Centrex single line:	$2.75
Centrex 2–8 lines:	$0.72
Centrex 9+ lines:	$0.31
Residential single line:	$0.53
Residential additional lines:	$1.50

your company pays the long-distance expenses at some of your employee's residences, your long-distance carrier is probably billing you the higher business line rate of $2.75 instead of the residential line rate of $0.53.

Another error with the PICC has to do with trunks. The PICC is to be charged for all access lines. Customers that use direct inward dial (DID) service are often billed too many PICCs. A typical customer may have 200 individual DID numbers but only 20 DID access lines. Long-distance carriers are prone to charge $2.75 for each of the 200 numbers, when they should only charge for the 20 access lines.

Tax problems

Long-distance bills are subject to the 3% federal excise tax, state taxes, and local taxes. According to Internal Revenue Tax Code Section 4253 (f) [1], some organizations are exempt from the federal excise tax [see Figure 11.4(b)] on "WATS and WATS-like services." No one uses WATS services anymore, but long distance is a WATS-like service and exempt organizations should not pay the federal excise tax on their long distance. Some exempt organizations are:

- Common carriers, such as trucking, shipping, and airlines;
- Certain news agencies;
- Certain government agencies;
- Nonprofit schools;
- Nonprofit hospitals.

Many of these organizations are unaware that they should not pay this tax. If your organization is exempt, provide the proper credentials to your carrier and it will remove the taxes from your bill. Some carriers will issue a refund if they feel the error is their fault, but most carriers will require you to secure the refund directly from the IRS. Trucking companies that want to avoid this tax should apply for a common carrier certificate. Exemptions from state taxes vary from state to state, and each situation must be researched individually.

Call detail

The last section of the bill [see Figures 11.4(b–c)] shows the detail of each call made during the billing period. The only thing unusual about the call detail is that the calls are being billed in full-minute increments. The first call is shown as a 1.0-minute call. The actual duration of the call may have been anywhere from 1 second to 60 seconds. Table 7.1 shows the true cost difference between full-minute billing and incremental billing.

Reference

[1]   Department of the Treasury, Internal Revenue Service, Publication 510, Revised March 2001, "Excise Taxes for 2001."

12

Long-distance pricing

When a consumer purchases a product such as a television, he usually knows the exact price prior to the purchase. If the price is unknown, no one will buy it. Long-distance pricing does not follow these rules of conventional wisdom. Most customers do not accurately know their long-distance rates. The actual rates that show up on the bill are rarely the same rates that were promised by the sales representative. This is one of the most confusing discrepancies in the entire telecommunications industry.

Some argue that long-distance pricing is confusing because carriers do not want customers to be able to calculate the rates. Probably closer to the truth is the simple fact that carriers use a complex system of discounting to calculate rates. Long-distance pricing is based on this formula:

$$\text{gross rates} - \text{discount} = \text{net rates}$$

The net rate the customer actually pays is based on a certain discount off the gross rate listed in the tariff. Table 12.1 shows examples of how long-distance rates are calculated.

Table 12.1   Long-Distance Rates After Discounts Are Applied

Call Type	Tariff Rate (Gross Rate)	Discount	Net Rate
Intralata	$0.12	10%	$0.108
Intrastate	$0.13	10%	$0.117
Interstate	$0.15	10%	$0.135
Intralata	$0.12	20%	$0.096
Intrastate	$0.13	20%	$0.104
Interstate	$0.15	20%	$0.120
Intralata	$0.12	30%	$0.084
Intrastate	$0.13	30%	$0.091
Interstate	$0.15	30%	$0.105

Tariffs

Tariffs are filed with both the state and federal government. Tariffs governing interstate services are filed with the FCC, while tariffs governing intrastate services are filed with the state's Public Utilities Commission (PUC). Telecom tariffs contain information on services offered, terms and conditions, and pricing. When a customer signs up for a new contract with a long-distance carrier, the contract always refers back to the tariff, and many of the specifics, such as price, are not documented in the contract itself.

The following quote is from an AT&T contract that illustrates that the rates are not disclosed in the actual contract but listed in a tariff. The two-page contract was offered to a customer with a 24-month term commitment and a $1,000 gross monthly revenue commitment. This contract is probably the most common long-distance contract in force today, and it serves as an accurate representation of the industry as a whole.

The service and pricing plan you have selected will be governed by the rates and terms and conditions in the appropriate AT&T tariffs as may be modified from time to time … AT&T reserves the right to increase from time to time the rates for the services provided under this tariff, regardless of any provisions in this tariff that would otherwise stabilize rates or limit rate increases… (AT&T's Business Service Simply Better Pricing Option Term Plan Agreement).

The contract quote reveals that rates are not specifically listed; instead, the customer is referred back to the "rates and terms and conditions in the appropriate AT&T tariffs." The sad news is that the average customer never sees the tariff. In many cases, the actual phone company sales representative who is offering the contract has never viewed the tariff either and is only familiar with the sales literature, proposals, contracts, and billing. I have negotiated more than 200 long-distance contracts. In only one case was the actual tariff provided to the customer, and in that instance, it was not because of a pricing issue.

Rate increases

Another problem revealed in the contract quote is the potential for rate increases. A few times each year, the major long-distance carriers increase the gross rates listed in their tariff. Over the past decade, a typical rate increase is 3% per quarter. Under this system, a customer's rates will have increased by more than 30% over a 2-year period.

For example, Acme Manufacturing spends $10,000 per month in long distance. Its current contract is about to expire, so the company's AT&T account executive offers a new contract. The proposal shows that the new pricing plan will drop Acme's monthly billing to $8,000. Acme, excited about saving $2,000 per month, signs a new contract with AT&T, which explains that the only way to get these great prices is with a 3-year agreement.

Every quarter, AT&T implements a 3% to 5% rate increase, and most customers never notice. If a customer notices the increase, he figures it is because employees are making more calls than before. After 2 years, Acme's bill is back up to $10,000, thanks to the rate increases.

AT&T then informs the customer of a new pricing plan that will cut the bill by $2,000 per month. The account executive says AT&T will be happy to void the current contract as long as it is replaced with a new 3-year contract. Rather than pay an extra $24,000 over the course of the next 12 months, the customer signs the new contract. And, you guessed it, every quarter the rates are incrementally raised, starting the cycle all over again.

The only way out of such a cycle is to bite the bullet and complete the third year of the contract. At that time, the customer will have maximum leverage to negotiate a better plan with the current carrier or may then switch to a lower cost carrier that guarantees its rates, such as Qwest, McLeod USA, and a host of resellers.

Carriers do not guarantee their rates because they fear certain market forces that could affect their revenues, such as inflation and new technologies. Ten cents a minute for long-distance calling is used as a benchmark today. If all the carriers guaranteed to charge their customers only 10 cents a minute, inflation would eat away the carriers' profits. New technologies such as free voice calls over the Internet also threaten a carrier's revenue potential.

The average phone company executives are eminently more concerned with pleasing their stockholders than they are with pleasing their customers. "Creating stockholder wealth" is the mantra for most companies. Sometimes the only way to take care of stockholders is to neglect customers.

Long-distance pricing

Even if your long-distance bill contains no errors, the pricing still makes the billing difficult to understand. Numerous factors, such as the following, affect the rates of a long-distance call:

- The termination point of the call: Is it intralata, intrastate, interstate, or international?

- The time of the call: Is it peak or off-peak?

- Whether or not the call is outbound, inbound, or calling card;

- Whether or not the call is switched or dedicated;

- Whether or not a virtual private network is in place.

Intralata, intrastate, interstate, and international calling

Long-distance bills usually separate the traffic into intralata, intrastate, interstate, and international calling. Long-distance carriers' international and interstate rates are listed in the tariffs they file with the FCC. Intrastate and intralata rates are listed in the tariffs filed with the state PUCs.

The interstate and international rates on one calling plan will be the same for all of a business' locations. For example, a business with locations in Illinois and Maine will pay the same rate for interstate calls at both locations. The intrastate rates, however, will be different. Intrastate pricing is governed by the tariff filed with the PUC in that state. The carriers set these rates based on the economic, competitive, and regulatory influences in each state. Intrastate rates in Illinois are about $0.08 a minute, while in Maine they may be as much as $0.30.

Intralata calls on a long-distance bill will have their own rate. The local carrier normally carries these calls, but many businesses have moved this traffic to their long-distance providers. Many long-distance carriers use the same rate for intrastate and intralata pricing, but the traffic is usually still separated on the actual phone bill.

In addition to having long-distance traffic itemized by interstate, intrastate, and intralata, the phone bill will also have a section for international calling. Calculating the true cost per minute for international calls is difficult, because a different rate is used for each country, but the bill combines all the international calling together. To effectively check international rates, you must spot-check individual calls in the bill's call detail section.

Call rounding

Long-distance rates may be whole numbers, such as 10 cents per minute, but more often they are expressed as fractional numbers such as 10.5 cents per minute. When customers double-check the rates on their long-distance bill, the rates are usually a little higher than the quoted rate.

Table 12.2 shows an example of call rounding. The customer was promised $0.069 per minute but is actually paying $0.071 per minute. On a large account, this 3% differential may be significant. If you have a legitimate error on your account, beware that your account executive may say, "The rates are a little high due to call rounding." This can only be true if the differential is less than a penny. If the difference is more than a penny, you probably have a different error on your account.

Peak or off-peak

Long-distance carriers offer lower pricing for off-peak calling to encourage callers to wait until the evenings. This makes more room on their network during peak times. Carriers are racing to increase their network capacity to keep pace with the fast growth of call volumes. At certain peak times, such as Thanksgiving Day, the majority of the public-switched network is in use, so many callers are unable to complete their calls. Lower off-peak rates should keep this from happening on normal working days.

Table 12.2   Call Rounding

Call Duration	Quoted Rate	Call Cost	Actual Cost on Phone Bill
30 seconds	$0.069	$0.035	$0.04
54 seconds	$0.069	$0.062	$0.07
60 seconds	$0.069	$0.069	$0.07
90 seconds	$0.069	$0.104	$0.11
2 minutes	$0.069	$0.138	$0.14
2 min., 30 sec.	$0.069	$0.173	$0.18
3 minutes	$0.069	$0.207	$0.21
5 minutes	$0.069	$0.345	$0.35
16.4 minutes		$1.132	$1.17
Quoted rate per minute:			$0.069
Actual rate per minute:			$0.071
Difference:			3.28%

Save money using off-peak calling

It is impractical for most businesses to shift their calling to the evening hours to take advantage of lower off-peak long-distance rates. A business that transfers computer data using modems and dial-up long-distance calling may be able to postpone these calls until the evening off-peak hours and save money.

A more practical suggestion is to compare the off-peak calling time offered by different carriers. If one carrier's off-peak time starts at 7 P.M. instead of your current carrier's 8 P.M. start time, maybe you should switch carriers. Telemarketing call centers that operate in the evenings can definitely profit from this suggestion.

Outbound long distance

Long-distance calls are processed through the long-distance carrier's network differently, based on whether or not the call type is outbound, inbound, or calling card. Because each call type uses different telephone company resources, the rates differ. When a carrier sets its rates, it has to consider the cost of access at the point of origination, the cost of transporting the call across long-distance lines, and the cost of access at the termination point. Figure 12.1 shows the different cost elements of a long-distance call.

In Figure 12.1, Jerry in Dallas, Texas, pays $0.12 a minute to call Linda in Atlanta, Georgia. His long-distance carrier does not own the physical phone lines from Jerry's house to Linda's house; it only owns the lines connecting the central offices. Lacking an end-to-end network, it must pay access fees to the local carriers on both ends for the use of the line. Access fees are between $0.02 and $0.04 per minute. Long-distance carriers argued for years that the access fees paid to local carriers are inflated and should be reduced. In this example, Sprint pays $0.06 in access fees and keeps the remaining $0.06.

One-time charges

When you make changes to your long-distance account, beware of one-time charges. These charges are often listed on the bill as "set-up charges" or "installation charges." Even if the amount of the charges is correct, carriers can almost always waive one-time charges. They are not always willing to waive these charges, but they are almost always capable of waiving the charges.

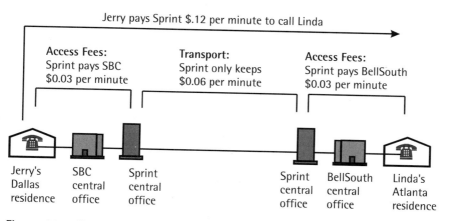

Figure 12.1 The cost of a long-distance call has three parts: access on the point of origination, transport, and access at the point of termination.

Because one-time charges are manually entered into the billing computers, the chance for human error is great. A manufacturer in the Midwest recently experienced a significant billing error with its carrier. The company added T-1 service in its domestic facility and at one of its Latin American facilities. The associated one-time charges should have been $1,060. These charges were never quoted to the company in advance because it routinely adds service at its various facilities, and the company trusted that the carrier would always bill it correctly.

When the company received its bill from the carrier, the charge was $106,000. The amazing part of the story is that the customer paid the bill and only months later began to question the charges. Its regular monthly bill was over $100,000 each month, and the extra $106,000 was not significant enough to immediately draw attention. When the company first questioned the carrier, the carrier's representative simply explained that the charge was a one-time charge for installation of the T-1 in Latin America, and that charges in Latin America are higher than they are domestically. After months of research, and hiring a consultant, the puzzle was finally solved.

One of the carrier's representatives explained that the overbilling was due to a simple data entry error. The person typing in the order accidentally typed in $106,000 instead of $1,060. Once the carrier admitted its error, it put a refund credit on the customer's next invoice.

13

Long-distance contract discounts

Long-distance rates are determined by applying a discount to the gross rate. The discount amount and the way it is applied differ between carriers. Each carrier offers multiple rate plans with varying discounts. Discount amounts even vary from one customer to the next. Most customers are content with their current rates until they become aware that lower pricing is available. Long-distance profit margins are high, which leaves plenty of room for customers to negotiate.

Volume and term commitments

The main factors that determine customers' discount amounts are the volume and term commitments in their long-distance contract. In return for the customer's promise to spend a certain amount for an extended period of time, the carrier offers a discount. The greater the volume and the longer the time, the greater the discount. Table 13.1 shows a typical discount structure used by long-distance carriers.

Most volume agreements specify the amount of net dollars spent each month. Net dollars are the actual dollars spent, not the prediscounted gross

Table 13.1   Typical Long-Distance Contract Discount Structure

VOLUME COMMITMENT	TERM COMMITMENT	DISCOUNT
$2,000	1 year	10%
$2,000	2 years	15%
$2,000	3 years	20%
$5,000	1 year	15%
$5,000	2 years	20%
$5,000	3 years	25%
$10,000	1 year	20%
$10,000	2 years	25%
$10,000	3 years	30%
$15,000	1 year	25%
$15,000	2 years	30%
$15,000	3 years	35%

amount. AT&T's Uniplan contracts calculate the volume using gross dollars on a monthly basis. Some volume plans are calculated annually. It is very important for customers to know if their volume commitment is net or gross and if the volume is calculated monthly or annually.

One-rate discounts

As the market becomes more competitive, carriers want their discount structures to be less complex so they can more efficiently set up new accounts. They also want potential customers to be able to easily compare their offer with other offers. That is why many long-distance companies are switching to one-rate billing with a level discount amount for all services, such as 30% off long-distance, paging, and mobile phones. If a carrier is trying to win a company's long-distance business, the proposal is normally clear and easy to follow. The phone bills, however, are not as easy to understand.

Save money with volume agreements

A simple way to reduce your long-distance bill is to increase your volume commitment, which will result in a greater discount amount. Most businesses wisely undercommit to avoid a shortfall penalty, but if you have extra volume, you should consider increasing your volume commitment level.

Your carrier will prefer that you sign a new contract with the increased discount, but you should first press the carrier to modify your existing agreement. If the carrier is inflexible, and you are not comfortable with a new agreement, you can move your "overflow" traffic to another carrier with lower rates. This will definitely get your carrier's attention. Many businesses use multiple carriers so their carriers never take them for granted. It is amazing how the level of customer service increases when a customer uses more than one carrier.

Save money with automatic discount upgrades

In many of its contracts, Qwest has a built-in clause to automatically increase a customer's discount if its volume hits the next highest level. For example, a small tax accounting firm committed to $2,000 per month with Qwest and received a 35% discount. From January through April, the firm's call volume doubled. In April, the bill passed the $4,000 mark, which is the next higher volume commitment level. Qwest automatically increased the discount to 40% for that month only. In May, the bill volume decreased again and the discount was back to 35%.

When is the true-up?

It is vital to understand how the actual long-distance usage will be reconciled against the contract's volume agreement. Long-distance accounts experience a true-up either monthly or annually. With a monthly true-up, the customer is required to bill at least his volume commitment each month. If he falls short, the carrier will add the difference to the bill. Annual commitments true-up the account at the end of the contract year. Figure 13.1 shows an example of a monthly true-up from a Telephone Company D bill.

This concept of the volume commitment true-up procedure is best illustrated with an example. Two brothers, Terry and Tony, each own their own summer resort. The business is seasonal; they rarely use the phone in winter. In the slow months, their long-distance billing is only $500 per month, while in the busy months, their billing rises to $1,500 per month. Table 13.2 shows a comparison of the billing for both brothers.

In January, Terry signs a new long-distance agreement that specifies a $12,000 annual net commitment. He understands that the true-up

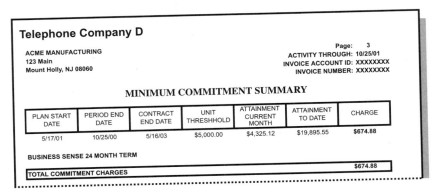

Figure 13.1 Sample of Telephone Company D's monthly true-up bill.

Table 13.2   Annual Versus Monthly Commitments

	$12,000 ANNUAL CONTRACT		$1,000 MONTHLY CONTRACT	
	Actual Usage	Monthly Billing	Actual Usage	Monthly Billing
January	$500	$500	$500	$1,000
February	$600	$600	$600	$1,000
March	$700	$700	$700	$1,000
April	$1,000	$1,000	$1,000	$1,000
May	$1,250	$1,250	$1,250	$1,250
June	$1,500	$1,500	$1,500	$1,500
July	$1,500	$1,500	$1,500	$1,500
August	$1,750	$1,750	$1,750	$1,750
September	$1,250	$1,250	$1,250	$1,250
October	$750	$750	$750	$1,000
November	$700	$700	$700	$1,000
December	$500	$500	$500	$1,000
Total:	$12,000	$12,000	$12,000	$14,250
Difference:				$2,250

*Additional charges are added to the account to meet the minimum commitment.

will happen at the end of the contract year in December. Tony follows his brother's lead and signs a similar agreement, but Tony's agreement specifies a $1,000 monthly net commitment. Tony does not read the fine print and is unaware that his account will experience a monthly true-up. Over the year, they used the same amount of long distance and have the same rates, but Tony ends up paying more than his wiser brother. At the end of the year, they compared their bills and found that Tony spent $2,250 more than his brother.

Usually, the true-up happens in the same period expressed with the volume commitment. A monthly commitment of $1,000 is reconciled each month. An annual commitment of $120,000 is reconciled at the end of each contract year. These guidelines hold true in all cases but one. The major exception to this rule is with AT&T's Uniplan billing. Uniplan volume commitments are expressed monthly, but the true-up happens at the end of the contract year. Therefore, a seasonal business is not penalized in its slow months.

Save money by avoiding shortfall penalties

If you are in a shortfall situation, you should contact your carrier immediately. Shortfall revenue is gladly accepted by carriers, but if the customer asks the carrier for relief, the carrier will normally negotiate an alternative. The key is to proactively address the situation before the shortfall charge is billed. The volume commitment can normally be reduced to the next lower level without having to sign a whole new agreement. Some of the discounts may be forfeited, however.

If the shortfall amount has already been billed, it is difficult for the carrier to simply waive the charges and reduce the volume commitment. Usually, the carrier will only waive the billed shortfall if the customer is willing to sign a new agreement with a new term commitment. I have seen customers in their last few months of a 3-year contract experience a shortfall and the only cost-effective way to avoid paying the shortfall is by signing a new 3-year contract. However, in this situation, the customer has little leverage and ends up paying high rates.

Contract value

Contract value is how carriers calculate how much money each customer is worth. Contract value is calculated by multiplying your monthly volume commitment by the number of months remaining on your term. For example, a customer at the beginning of a $1,000 per month, 12-month agreement has a contract value of $12,000. The same customer 10 months later is only worth $2,000 to the carrier. By studying the contract value of the entire customer base, long-distance company financial analysts can predict future revenues.

A customer in a shortfall situation should be aware that his carrier uses the contract value principle to guide him during negotiations. A wise customer considers this same principle when negotiating with her carrier. To clear up a shortfall, your carrier will always require you to increase your contract value. So a customer facing a $10,000 shortfall penalty must sign a new contract that promises the carrier at least $10,000 in future revenue.

Term agreements

Carriers normally offer 12-, 24-, and 36-month term agreements. The longer a customer will commit to a carrier, the greater discount the carrier will offer. The combination of the term agreement and volume commitment establish the discount amount. Table 13.1 illustrates how a typical long-distance carrier structures its discounts.

On national accounts, carriers will normally push for an even longer term agreement, such as 48- or 60-month agreements. Ironically, the longer your term agreement, the less attention you get from your carrier. The carrier knows that they have no risk of losing your business in the short term, so they focus their attention on their more volatile customers.

Save money with term agreements

Increasing your term commitment increases your discount amount. Carriers push the 36-month term agreement because they want to count on the customer's revenue for as long a period as possible. The pricing difference between 24- and 36-month agreements is often negligible, so the customer should choose the shorter-term commitment.

Many account executives offer a new 36-month term agreement as their standard offer. If the customer is a savvy negotiator, he can often secure the same pricing on a 12-month agreement. A rule that guides many consultants is to simply reject the long-distance carrier's first proposal. Consultants know from experience that account executives rarely offer the best pricing with their first proposal. In this way, long-distance contract negotiation differs little from the negotiation done while buying a car.

In some cases, a customer's current long-distance contract may be amended to increase the term. In other words, another year can be added to the agreement without requiring a new contract. In most cases, however, increasing the term agreement to secure lower discounts is normally done during the initial contract negotiation.

Some small areas of the country are not yet "equal access." That means that customers in those areas can only choose AT&T as their long-distance carrier. For example, a Midwest aluminum siding company is located in a rural area surrounded by farm fields. Its facility uses more than $5,000 per month in long distance, but the company can only use AT&T. None of the other carriers have built a network out to this remote area. This business might as well sign the maximum term agreement available because it has no other choice of carrier. At least with a long-term agreement, it can secure the lowest rates available through AT&T.

A business that receives specialized services from one carrier that cannot be duplicated by another carrier should also sign a long-term agreement with its carrier. An oil prospecting business located in Texas spends more than $10,000 per month in long distance. Most of the billing is from calls made by field representatives who are in remote areas of the Middle East. The field representatives use calling cards for their calls. The telecommunications infrastructure is underdeveloped in the oilfields of the Middle East, and only AT&T can satisfactorily provide this service.

Other carriers would like to earn the company's business, but the company cannot afford the risks associated with trying a new carrier. This business has no reason to not sign a long-term agreement with its current carrier.

Save money with association discounts

AT&T's Profit By Association (PBA) plan gave it a highly effective marketing tool. A customer who was a member of one of many associations, such as AAA, received an additional 5% discount. The long list of approved associations allowed almost every business to qualify for the PBA discount. The plan was very successful in drumming up new business for AT&T, especially when sales representatives set up a new PBA through the local chamber of commerce.

If your business has no membership in a participating association, consider joining one if for no other reason than to cut your long-distance bill by 5%. One enterprising AT&T account executive in Illinois created his own Secretary's Association. Any business that has a secretary can join the association by paying only a $10 annual membership fee. Because every company has a secretary, the sales representative was able to offer this additional discount to almost all of his prospects. Similar association plans are available with other carriers.

Save money with international discounts

Enrolling in an international discount plan can be an effective way to cut your long-distance bill. These plans give an additional discount on international calls to one or more countries of the customer's choice. AT&T's plan, called the Favorite Nation Option, gives the customer an additional 10% discount on calls to a single country. Other carriers offer a discount on a group of countries, such as Latin America or the Far East.

Save money with referral programs

From time to time, long-distance carriers may offer a referral discount plan. Before LCI merged with Qwest, it offered a Goose Eggs referral program. This program gave a company an additional 2% discount for every company it referred that switched its long distance to LCI. The goal was to refer 50 customers, which would result in a 100% discount. The customer would then receive his bill every month with "goose eggs" in the bill's amount due section. Other referral programs apply discounts based on the bill volume of the company referred. So if the new customer spends $1,000 per month, the referring customer sees a $50 credit on her bill each month.

Points programs

Some carriers have created their own points programs similar to the airlines' frequent flier mileage programs. For the past few years, Sprint's Callers' Plus Points program has been very successful. For each dollar spent on long distance, a customer earns one Callers' Plus point. Every 50 points can be applied as a $1 invoice credit, or the points can be redeemed for merchandise from Sprint's catalog. The catalog contains items such as televisions, hotel nights, and office supplies. The catalog is often an attractive option for a company controller, because merchandise can be secured without using money from a budget.

Participating in this program may be a hassle, but the additional 2% bill credit may make it worthwhile.

Missing discounts

The most common error on long-distance bills is a missing discount. The discount amount may be incorrect or missing altogether. Incorrect discounts often happen because an overzealous sales representative offered higher discounts than he was allowed to offer. When the order reached the carrier's data entry staff, they lowered the discounts to the correct amount. If this happens, a customer should present the initial proposal to the sales representative and sales manager and demand that the billing be corrected.

Missing discounts on subaccounts

Companies with multilocation billing often experience billing errors, especially with Sprint. An East Coast real estate company had offices in three cities. The company switched to Sprint and had all three locations bill on the same master account. Sprint set up the account and gave each location the 35% discount the customer negotiated. The 35% discount is based on standard tariff discounts and an additional 10% custom discount. The customer examined the bills regularly and did not detect any errors.

Later, the firm opened two more locations and had Sprint add them to the account. When Sprint set up the two new subaccounts, it failed to implement the custom discounts, so those locations ended up paying higher rates. I have seen this error on at least a dozen Sprint accounts. After being notified, Sprint usually corrects the problem quickly and issues a refund.

National accounts

A basic rule of economics is that the more a customer spends, the better pricing he receives. This principle is at the heart of long-distance carriers' national account pricing models. A large national company with multiple locations, such as Holiday Inn, can aggregate all of its long-distance volume under one national contract with a single long-distance carrier.

Even though each individual hotel spends only $2,000 to $5,000 per month in long distance, collectively all the properties spend more than $1 million per month. Each location individually is not enough to turn carrier's heads, but the opportunity to win the entire corporation's long-distance business is enough to make carriers bring their best offers. Each hotel property will then pay long-distance interstate rates as low as $0.03 a minute, while an independent hotel may pay $0.10 per minute.

Ironically, long-distance carriers usually do not bring their best offers when bidding for a national account such as Holiday Inn. Instead, they steer the negotiations around a separate topic such as network reliability, network features, or billing features. These items are of the utmost importance to a national business, but differences between the carrier offerings are minimal. As stated before, long-distance service is more of a commodity today. Besides the customer service provided by the individual account team, the only thing separating one carrier from another is price.

Long-distance carrier national account teams are some of the best negotiators in the business world today. This statement is based on the dozens of national accounts with which I have worked. In almost half of the cases reviewed, I found that residential users spending $25 per month were able to negotiate lower rates than these Fortune 500 companies.

One Fortune 500 company was content paying $0.25 cents a minute for dedicated interstate long distance, because it was "too much hassle to have the long-distance carrier work on better rates." The customer cited that the carrier was too busy with her numerous requests for the home office to worry about a remote site. This practice cost the company thousands of dollars each month in unnecessary overbilling, but the controller at that site refused to change anything.

Autonomy: Who is in charge?

Some national companies specify which vendors their remote locations may use. For example, Holiday Inn's corporate office may mandate that each hotel use Sprint for long distance. This arrangement can prove very frustrating in the case of a franchise, where an individual owns and operates his own hotel under the auspices of Holiday Inn. With other national businesses, each location is autonomous; it can choose its own long-distance carrier. In this instance, the location should compare the difference in rates between its current plan and the national plan negotiated by the home office. The national account pricing may be very aggressive, and it may make economic sense to switch the long distance to the national plan.

When a remote location joins the national account of the home office, the phone bills might be consolidated. The remote site's usage will show up on the home office's bill. This creates additional internal accounting work for the customer.

The home office gets all the goodies

Most of the larger carriers can create a billing platform that allows the corporate office to be subsidized by the remote locations. This is accomplished by having every location pay the same rates, but the home office receives a disproportionate discount amount. So, for example, with a national chain of franchise hotels, each hotel may pay long-distance rates of $0.10 a minute, but the home office only pays $0.02 a minute. In extreme cases, the remote hotels may pay $0.12 per minute, while the home office has free long distance or is paid a commission check each month. Such an arrangement can be costly for a business, unless, of course, you are the home office.

Another problem with national account billing arrangements is that the home office "gets all the goodies," as one consultant said. National account teams are encouraged to give premiums to their clients. If a national account bills more than a $100,000 a month in long distance, the account team will lavish the home office with gifts such as free prepaid calling cards, golf outings, free lunches, and vacations. These premiums have proved to be very instrumental in helping the account team retain an account. Such premiums have also influenced many customers to choose the wrong long distance for their corporation. The opportunity for personal gain sometimes outweighs a person's ability to objectively manage his organization's expenses.

Move away from a national account

Because of the numerous pitfalls associated with national account billing plans previously mentioned, it might be beneficial to cancel the plan altogether. A national account that allows its locations to manage their own expenses can set a pricing benchmark and require all locations to negotiate their own long-distance contract as long as the rates are below the benchmark. This is not easy, however, because bills are confusing and rates are normally difficult to interpret.

Normally, the greatest cost reduction strategy with national accounts has to do with remote locations separating from the national account. If the remote locations are autonomous and can choose their own long-distance plan, they should calculate the rates they are paying on their current bill with the national account and negotiate a better plan on their own if they are free to jump ship.

14

Dedicated service

This chapter will explain the benefits of using T-1 service to reduce your long-distance expenses. T-1 service can provide medium-size businesses with very significant savings for long distance. The objective of this chapter is to offer the first strategic steps to determine whether or not your business should use T-1 service. If you decide to pursue this option, you will then need to consult your telephone equipment vendor and long-distance carrier to find out exactly what is required to implement T-1 service from a technical viewpoint.

Is it switched or dedicated?

A switched connection is a temporary connection made between two points by passing through a switching device such as a phone company central office. A dedicated connection is a permanent connection made between two points.

Long-distance calls are either billed as switched or dedicated. Most phone calls are switched. All residential calls are switched calls. Figure 14.1 shows how a switched long-distance call works. When a caller dials a

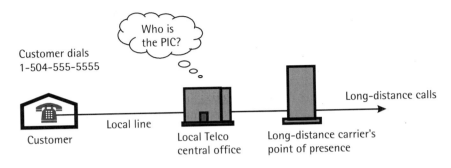

Figure 14.1 Switched long-distance call.

long-distance number, the local carrier's central office interprets the "1 + area code" that was dialed and then switches the call to the long-distance carrier's nearest point of presence. The long-distance company then carries the call across its network. On the long-distance carrier's bill, this call will show up under the heading "Switched Long Distance."

The term "switched" refers to the fact that the first leg of the call was switched at the local carrier's central office. This also means the long-distance carrier is paying access fees to the local carrier for this call. Figure 12.1 shows how the access fees are charged on a long-distance call. For this reason, switched long-distance rates are normally 3 to 4 cents higher than dedicated long-distance rates. If the access fees could be avoided, then the long-distance rates the customer pays should theoretically be 3 to 4 cents lower. This is precisely the reason why dedicated long-distance rates are lower than switched long-distance rates.

Dedicated long-distance service uses a T-1 circuit from the customer's premise that connects directly through the long-distance carrier's point of presence. This dedicated line carries all of the voice long-distance calls. The calls bypass the local carrier's network, so the long-distance carrier does not pay access fees for these calls, which results in lower rates for the customer. Figure 14.2 illustrates how a dedicated long-distance call works.

Although the long-distance rates decrease with dedicated service, the customer will have the added monthly expense for the T-1 circuit. Even though the local carrier provides the physical circuit, the T-1 is usually billed by the long-distance carrier. The long-distance carrier will charge a fee for securing the T-1 facility from the local telephone company. This fee is usually called "access coordination" and costs about $85 per month. T-1

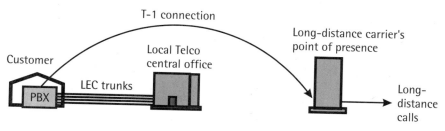

Figure 14.2 Dedicated long-distance call.

monthly costs range from $250 to $1,200 per month, depending on the mileage from the central office and the discount amount. The greater the mileage, the greater the cost.

T-1 installation

The one-time initial T-1 installation typically costs $1,000. The long-distance carrier will normally waive the installation cost if the customer has some negotiating leverage. If you are a new customer with the carrier, it will almost always waive the cost of installation because it is eager to win new business. If you have had any recent billing errors or service issues, carriers will almost always consider your inconvenience and waive these charges.

Beware, however, that your long-distance carrier representative normally only quotes the installation costs associated with the T-1 line. He probably does not have the expertise to determine whether or not your telephone equipment will be compatible with that T-1 line. Prior to placing any orders with your long-distance carrier, it is imperative that you consult with your equipment vendor.

In some cases, the telephone equipment may need costly upgrades that could make the project cost prohibitive. If the equipment only needs minor upgrades, the long-distance carrier may issue an invoice credit to cover the expense of the new equipment.

Save money by moving to dedicated service

When considering moving from switched long-distance service to dedicated service, a simple cost comparison must be done. The up-front installation costs are normally factored in the first year's numbers. The soft-dollar expenses of the additional time it will take to manage the conversion should also be considered, even though the financial impact is difficult to quantify. The potential cost savings of dedicated service are best illustrated in the following example.

Smith Designs is a young company that sells home decorations through a catalog and a Web site. The business has grown significantly over the past few years, and now the ordering and customer service is handled by a small call center staffed by 20 employees. Smith has always used switched long distance, but the long-distance representative is now proposing dedicated T-1 service. Table 14.1 shows the cost comparison.

Should we install a T-1?

When considering installing a T-1, you should ask your long-distance provider and your equipment provider the following questions:

- ◆ What equipment upgrades, if any, does your phone system require?
- ◆ What are the costs of these upgrades?
- ◆ Will the long-distance carrier cover this expense?
- ◆ What is the cost of installation for the T-1?

Table14.1   Switched Versus Dedicated Rates

		SWITCHED		DEDICATED	
Current Minutes		CPM*	Current Cost	New CPM	New Cost
Outbound					
Intralata	3,164.0	$0.080	$253.12	$0.055	$174.02
Intrastate	5,467.4	$0.090	$492.07	$0.060	$328.04
Interstate	16,546.3	$0.105	$1,737.36	$0.070	$1,158.24
Inbound					
Intralata	3,548.2	$0.080	$283.86	$0.055	$195.15
Intrastate	5,453.0	$0.090	$490.77	$0.060	$327.18
Interstate	13,548.4	$0.105	$1,422.58	$0.070	$948.39
Monthly T-1 cost					$450.00
Total:			$4,679.76		$3,581.02
Monthly difference:					$1,098.73
Annual difference:					$13,184.78

*CPM = cost per minute

* Will the long-distance carrier waive this expense?

* What are the new domestic rates? International rates?

* How long will the installation take?

* How soon can the installation be scheduled?

* How much time will pass between the signing of the contract and the T-1 installation?

* Can we schedule the conversion to take place on the weekend?

* How will the conversion be tested?

Save money by removing dedicated service

T-1 monthly recurring costs have declined over the past few years. More and more customers are using dedicated long-distance service because of the tremendous opportunity to reduce costs. Some businesses, however, are doing just the opposite and are canceling their T-1s to cut costs. This is especially true for manufacturers that are closing facilities. When a facility is closed, a skeleton crew of workers remains at the old site for a year or two. They will make fewer long-distance calls and no longer need their T-1. They can instead allow their long-distance calls to be routed across regular local lines.

Save money by moving switched loose lines to dedicated

A common long-distance inefficiency is having switched long distance at a location that has dedicated service. A printing company routed its AT&T long distance across a T-1 for almost 10 years. When it ordered a new 800 number, AT&T's customer service representative overlooked that the 800 number should be routed across the T-1, so the 800 number rang in on ordinary local telephone lines. Once the problem was discovered, AT&T agreed to redirect the 800 number to ring in through the T-1. This cut the company's cost for these calls from $0.10 a minute to $0.06 a minute.

Long-distance virtual private networks

Some businesses transmit data and voice calls across their own private networks such as on a college campus. Because all of the lines on campus are owned and maintained by the college, the college does not have to pay monthly phone bills for these calls. When someone at the college calls someone off campus, the public-switched network is used, and this usage is itemized on a phone bill.

Few organizations can afford to build their own private network, but they can still have some of the benefits of a private network by using a virtual private network (VPN). VPNs use public telecommunications infrastructure, but the carriers often provide more secure, private connections than a normal phone call would experience.

VPNs are used more for data than for voice calls. In today's market, long-distance carriers will propose to a large client that the carrier be allowed to handle both voice and data traffic. Numerous technical issues must be clarified with the carrier if data traffic is to flow across the VPN. My objective is to explain how long-distance voice calls are affected by a VPN configuration.

VPNs connect all of a customer's major locations through the long distance carrier's lines. A location with dedicated T-1 service is an *on-network* site, while a location with switched service is an *off-network* site. Different rates apply to calls based on whether or not the call is on or off the network. Table 14.2 shows how long-distance rates are structured in a typical VPN.

All this may seem confusing, but it is relevant because it affects the long-distance pricing for a business that has a VPN billing arrangement with its long-distance carrier. Carriers sometimes make billing errors by classifying a T-1 location as an off-net location, which results in the wrong rate being charged.

Save money with tie lines

Numerous businesses connect their locations with dedicated phone lines so that their computers can transmit data files back and forth. Without a private line, the business would have to send the data across normal phone lines using modems on both ends. If the locations are not in the same

Table 14.2   VPN Rates

A call from an on-net location to an on-net location	$0.05 per minute
A call from an on-net location to an off-net location	$0.07 per minute
A call from an off-net location to an on-net location	$0.07 per minute
A call from an off-net location to an off-net location	$0.10 per minute

LATA, each call will be billed by the minute on the long-distance bill. If the call volume grows significantly, at some point it is more cost effective to pay for a dedicated connection. Private line pricing is based on bandwidth and mileage. To calculate the break-even point, simply compare the cost of a private line to the current cost of the dial-up calls.

A not-so-new trend in long distance is to migrate the voice long-distance calls across the same dedicated connection, as long as the connection can spare the extra bandwidth for the voice calls. In this scenario, the dedicated line is called a tie trunk because it connects two PBXs.

For example, a manufacturer in Maine made frequent long-distance calls to its office in Vermont. The company eventually automated its assembly line and had to share computer data between the two locations. Initially, the computers dialed each other and sent the data across normal phone lines. This became very expensive because the computers were calling each other throughout the day.

The telecom manager decided to install a T-1 line between the two locations to carry both voice and data long-distance calls (see Figure 14.3). This measure dramatically reduced the company's costs, but it did not stop there. The Maine location made a large number of intrastate

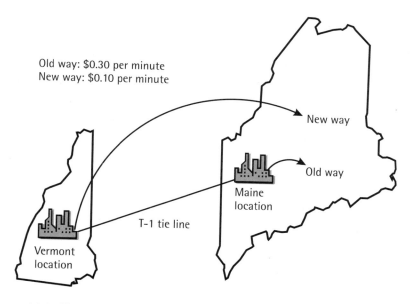

Figure 14.3 Tie lines.

long-distance calls to customers and suppliers within the state. Maine intrastate rates are the highest in the country. In fact, some international rates are cheaper than Maine intrastate rates even though the actual distance is much greater.

To reduce the cost of the Maine intrastate calling, the telecom manager programmed his PBX to route all Maine intrastate calls through the Vermont office first. In doing so, these calls would be billed at the low interstate rate instead of the higher Maine intrastate rate. The call delay for the added mileage is undetected by the end user, for it happens in milliseconds.

15

Data networking essentials

What is data networking?

Today, most companies, large and small, use data networking technology in their day-to-day business. An automobile parts manufacturer in Chicago sends electronic production reports to two separate manufacturing sites across two dedicated T-1 lines. A florist in San Diego uses an ISDN connection to the Internet to rapidly communicate with her suppliers and customers. The WAN of a Nebraska-based medical insurance company connects 29 different offices, uses 24 separate phone companies, and costs slightly more than $100,000 per month.

Each of these data networking examples requires both computer and telecom technology to work correctly. In all three cases, the companies use computer and telecommunications technicians to install and maintain the network. The expertise of these technicians is invaluable. They have spent years mastering leading-edge technologies that are perplexing to the average businessperson. In most organizations, however, it is a businessperson who is responsible for the data network. The most efficient, cost-effective data networks rely on the expertise and savvy of technicians and businesspeople.

Numerous resources that explain the inner workings of the technological aspects of a data network are available. The purpose of this book is to provide businesses with strategic advice on managing the expenses of voice, data, and wireless telecommunications services. This chapter offers an overview of data networking technology explained in layman's terms.

In its simplest definition, data networking is "two or more computers communicating over a medium." The communication may be considered local such as the connection of multiple computers across the inside wiring of a Chicago office building. Or the communication may be across a wide area such as a computer in Denver connecting to the Chicago office across telephone company lines. In the first example, the *medium* is the inside wiring in the Chicago office. In the second example, the medium is the telephone company phone lines.

LAN and WAN

LANs are normally wired with company-owned inside wiring. It costs the company nothing to transmit data across this medium because it owns the wiring. In the case of a WAN, a business transmits data across a long distance, or wide area, such as from Dallas to Chicago. WANs use the phone company's network, so the phone company bills the customer each month for this service. Because this is a cost management book, we will deal with the telecommunications offerings and billing associated with WANs, not LANs. Within this context, our definition of data networking is "two or more computers communicating over a telephone line."

The information revolution

The last 20 years can be viewed as a sort of information revolution. Like the Industrial Revolution that radically changed the world a century ago, the computer age has also ushered in a new era. Computers have impacted business, industry, employment trends, and consumer behavior. Numerous jobs are now obsolete due to computers, and an increasing number of people spend their working hours sitting at a computer workstation. And, like the Industrial Revolution, the information revolution has produced numerous consumer benefits.

While the end user is happily sending and receiving data, behind the scenes, the telephone companies are investing millions of dollars building up these networks. They, in turn, bill their customers for all this data traffic. In 2000, telecom carriers billed their data networking customers more than $7 billion.

Applications

Commercial use of data networking falls into two categories: *front-office applications* and *back-office applications*. A dime store's customers walk in the front door and its suppliers walk in the back door. Data networking applications are essentially the same. The front office handles *customers*, while the back office handles *suppliers*. Front-office applications concentrate on customer relationship management and include:

- Customer service;
- Point of sale systems;
- Call centers;
- Sales;
- E-commerce;
- Shipping.

Back-office applications concentrate on internal company business and external supplier relationships such as:

- Accounts payable;
- Inventory control;
- Accounts receivable;
- Logistics;
- Engineering;
- Manufacturing;
- General ledger accounting;
- Research and development;
- Human resources;
- Supply chain.

Distinguishing between front- and back-office applications is important to a network's stability and security. Front-office applications are more "mission critical" to the success of the business. Consequently, the data network that handles front-office applications should be less prone to glitches, downtime, and security breaches.

Back-office applications also need to be secure and reliable, but the business suffers more when front-office applications crash, so they have priority. Just like the five-and-dime store, a company loses more business when its front door is broken than if its back door is broken.

Network evolution

The methods for sending and transmitting data across telephone lines have changed over the past few decades. Data networks have evolved over a series of generations. Some of the more significant data network milestones are the mainframe computer, the personal computer, LANs, WANs, the Internet, intranets, and extranets. Many companies incorporate all of these technologies in their corporate data networks. IBM calls this combination of an organization's networks its "enterprise network."

Mainframe computer networking

In the beginning, mainframe computers handled all data processing. "Dumb" terminals displayed the results of the mainframe computer's work. Remote dumb terminals connected with the mainframe across expensive leased telephone lines.

For example, a national hotel chain used a mainframe at its home office to handle reservations. The dumb terminals at each hotel property connected to the mainframe for data processing and to access reservation records. The connection took place across telephone company leased lines. The phone companies "dedicated" the leased lines for the hotel chain's use only, so they were very costly, but there was no other way for the hotel to do business. It was too expensive to purchase a mainframe for each hotel to do data processing, and each hotel still needed access to the home office's reservation records.

Personal computers

The greatest distinction between mainframes and PCs is that PCs allow individuals to process their own data without the need to be connected to a mainframe computer. Using a modem, one PC can dial up another PC and share data. So, for example, the hotel chain could scrap its mainframe and use PCs at each location for data processing, but each hotel would still need to connect to the home office to handle reservations. Depending on the amount of traffic, a dial-up modem connection may be more cost effective than dedicated lines, but it would not be as reliable.

LANs

LANs are simply two or more computers connecting to each other in one location. For example, the network used by a small law firm consisting of a server, a printer, and five personal computers is considered a LAN. Figure 15.1 is an example of a LAN.

WANs, MANs, and VPNs

A WAN is two or more computing devices in separate locations connecting to each other across telephone lines. Figure 15.2 shows an example of a typical WAN. WANs connect with dedicated private lines, circuit switching, packet switching, or a combination of these services, all of which are further explained in Chapter 16.

Most WANs can also be called VPNs. A truly private network means that in addition to owning the computing equipment, *the user also owns the phone lines* that connect the computers. Some universities and corporate campuses have private networks. Building a physical telecommunications network is very difficult and very expensive; few organizations use private networks. Instead, they have private communications over lines owned by the phone company. This is not a private network; it is a "virtual" private network.

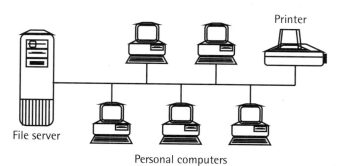

Figure 15.1 Typical LAN.

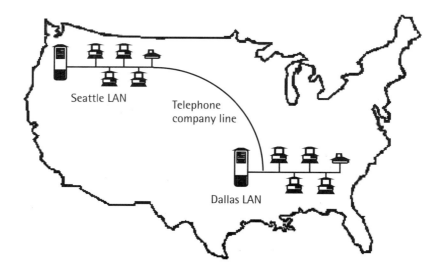

Figure 15.2 WAN.

When telecommunications professionals use the term VPN, they are usually referring to large businesses that transmit voice and data across a data network. A metropolitan-area network (MAN) simply refers to a WAN that is contained within one metropolitan area.

The Internet, intranets, and extranets

The Internet

The Internet was originally a private network developed to link universities, defense contractors, and the U.S. military. *Newton's Telecom Dictionary* correctly calls the Internet "the most important happening in the telecommunications industry since the transistor." Today, the Internet is the world's largest computer network. Businesses typically have two types of Internet-related telecommunications expenses: Web site hosting and Internet access.

Intranets

An intranet is a network using Internet software and standards. An intranet may or may not reside on the Internet. The portal to the intranets of most large companies can be seen on their Web sites under the heading "For Employees Only."

Extranets

The term extranet refers to a company making its own network available to consumers or other businesses across the Internet. Extranets are generally set up for electronic commerce or data exchange between a business and its suppliers or a business and its customers. Electronic commerce is the process of buying or selling something over the Internet. Buying a book from Amazon.com or booking airline tickets through Expedia.com are examples of e-commerce. A shipping company, such as UPS, that allows customers to track their packages on-line is another example of an extranet.

Most of the terms mentioned, such as LAN, WAN, MAN, intranet, and extranet, serve as conceptual terms and are not exact definitions.

Types of data

Three types of traffic travel on telecommunications networks: voice, data, and video. Voice traffic is two people talking on the phone; data traffic is two computers talking to each other. A remote manufacturing facility downloading computer files from a corporate headquarters office is an example of data traffic. Video conferencing is usually done in the context of a business meeting, or more likely, a university distance-learning class.

Until the 1970s, telephone networks primarily carried voice traffic only. Since then, the boom of computers has caused more and more data traffic to move across telephone lines. Around 2000, the industry experienced equilibrium with the amount of data traffic equaling the amount of voice traffic. Videoconferencing is viewed by most businesses as an inconvenient novelty and, as such, is not widely used today.

Convergence

When two or more of types of data are sent across the same connection, as is possible with innovations such as time division multiplexing and packet switching, this is called *convergence*. The different data types converge across the same telephone line.

Media

By definition, telecommunications is simply communication between two points across media. The childhood tin can phones connecting two tree houses use a string as the medium. Networks either use *physical* media, such as copper wire, coaxial cable, or optical fiber, or *broadcast* media, such as radio waves, microwaves, or satellite signals.

Media is not the concern of the average customer. Most customers do not care how high-tech their carrier's network is. As long as they can pick up the phone, make calls, and be billed a fair price, they could care less if the calls are on fiber-optic lines or 50-year-old copper wiring. But phone companies love to boast and brag about the superiority of their networks, and they are not above criticizing another carrier's network. After an AT&T network crash in the late 1990s, MCI quickly issued bold press releases, boasting "that would never happen with our network," but in the fall of 1999, MCI's frame relay network suffered a 10-day outage. The crash was infamous because it caused trading to be suspended at the Chicago Board of Trade, one of MCI's key customers.

The last mile

The greatest media concern for most customers is the last mile. The last mile is the cabling from the telephone company central office to the end user's premise (see Figure 15.3). Just as a chain is only as strong as its weakest link, a telecommunications network is only as fast as its slowest link. While phone companies deploy miles and miles of new fiber-optic cabling across the "backbone" of their network, the last mile is usually twisted-pair copper wiring that transmits an analog signal.

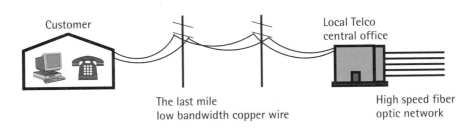

Figure 15.3 The last mile.

The limitations of the last mile greatly hinder the speed and efficiency of telecommunications. The world is full of people who surf the Internet with the latest high-speed modems but are frustrated because the service is too slow. New services such as integrated services digital network (ISDN), DSL, and cable modems offer faster connections, but until these services are widely offered, the last mile is still a giant speed bump in the middle of the information superhighway.

Bandwidth

Businesses and consumers alike want fast computer connections. A stockbroker who wants to quickly buy and sell stock electronically cannot tolerate a slow, unreliable connection. Numerous variables affect how fast the information is sent. The overall speed of the transmission is affected by the sending computer, the receiving computer, the networking equipment at the customer's premise, the phone company's networking equipment, the cabling at the customer's premise, and the last mile of cabling used in the carrier's network.

Telecommunications professionals use the term *bandwidth* to refer to the capacity, or speed, of a data transmission. Bandwidth simply refers to the amount of data that can be sent across the line. Bandwidth is like speed. The greater the bandwidth, the more data can be sent at one time.

If a stockbroker uses a low-bandwidth connection, for example, her orders may be placed a few seconds later, or even a few minutes later. This kind of variable is very costly in the high-speed, high-stakes world of Wall Street. Once the stockbroker decides to seek out a faster data connection, she may contact her phone company and upgrade to another service that provides more bandwidth.

Telephone companies offer a variety of data services today, such as dedicated private lines, ISDN, frame relay, ATM, and DSL. When selecting the most appropriate service, customers must consider the cost of the equipment, installation charges, monthly expenses, the service's user friendliness, and, most importantly, the service's bandwidth. The greater the bandwidth, the greater the price. Businesses must carefully choose the right amount of bandwidth for their application or they may be wasting money.

Most computers in use today are faster than the telecommunications networks to which they are connected. The telephone companies are engaged in massive campaigns to beef up their networks to meet demand.

The customer demand for bandwidth is also growing, and it looks like we will still have to wait a few years before there is enough bandwidth to go around for everybody.

Analog services send data by modifying the electromagnetic waves that travel along a copper telephone wire. Digital services send data across copper wiring with electric "on and off" signals. Across fiber-optic lines, digital services send data by flashing a light on and off. The faster the light is flashed, the more bandwidth the user has. As the signal travels across the line, telecom networks amplify the signal. During transmission, the signal fades and is distorted by outside "noise" (see Figure 15.4). Analog service cannot distinguish between the true signal and the noise, so it amplifies both. Digital service maintains a clean signal throughout the transmission.

Bandwidth is measured in bits per second (bps), kilobits per second (Kbps), or megabits per second (Mbps). The maximum bandwidth for a normal modem connecting across a normal telephone line is 56 Kbps. The most common bandwidths offered by telephone companies for data services are: 56 Kbps, 128 Kbps, 256 Kbps, 1.544 Mbps, and 44.736 Mbps. To illustrate the concept of bandwidth, Table 15.1 shows how much time it takes to send the complete manuscript of *Gone With the Wind* using various bandwidth amounts. Table 15.2 shows how many simultaneous phone calls each connection can handle and the approximate monthly cost of each service.

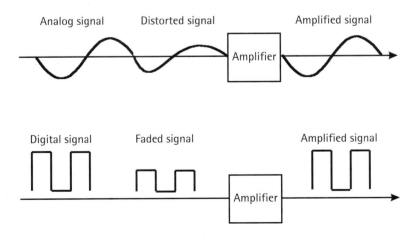

Figure 15.4 Analog and digital signals.

Table 15.1   Length of Time to Send the Text of
Gone With the Wind Across Telephone Lines

TYPICAL USER	CONNECTION SPEED	AMOUNT OF TIME TO TRANSMIT THE TEXT OF GONE WITH THE WIND
Home computer	14.4 Kbps	3 minutes, 8 seconds
	28.8 Kbps	1 minute, 38 seconds
	56.0 Kbps	49 seconds
Dedicated private line linking two business locations	Fractional T-1 256 Kbps	10 seconds
Access to a long-distance carrier Access to an Internet service provider Dedicated private line linking two business locations	T-1 1.54 Mbps	2.5 seconds
Large corporations, universities Telephone companies	T-3 44.7 Mbps	10 book copies per second
Large corporations, universities Telephone companies	ATM 622.0 Mbps	140 book copies per second
Telephone companies	SONET 2,500.0 Mbps	560 book copies per second

Table 15.2   Number of Simultaneous Phone Calls Each Service Can Handle

CONNECTION	BANDWIDTH	SIMULTANEOUS PHONE CALLS	MONTHLY COST
POTS line	Up to 56 Kbps	1	$35
T-1 service	1.54 Mbps	24	$300–$1,200
T-3 service	44.7 Mbps	672	$5,000–$10,000
ATM	622 Mbps	Over 9,000	Depends on wholesale pricing and mileage
SONET	2,500 Mbps	Over 35,000	

Many companies lease dedicated point-to-point telephone lines from their long-distance carrier to connect LANs at two different locations. The two sites can then share computer files. Businesses also lease dedicated lines from phone companies for high-speed connections to their Internet service provider (ISP). Businesses usually choose 56-Kbps, 128-Kbps, 256-Kbps, or 1.544-Mbps bandwidth for these types of applications. T-1 service has 1.544 Mbps of available bandwidth; 128 Kbps and 256 Kbps are often called "fractional T-1" bandwidth.

Although the service can be expensive, customers like T-1 service because it provides a lot of bandwidth. Phone companies like T-1 service because it only requires the amount of cabling that two ordinary phone lines would use. When used for long distance, T-1 service is divided into 24 channels so that 24 simultaneous phone conversations can be carried.

Some businesses split their T-1s between voice and data. A real estate company in Dallas, for example, uses 16 channels for outbound long-distance calling, 4 channels for inbound calling, and the remaining 4 channels for Internet access. All of these services are provided by the company's long-distance carrier.

What is a T-1?

When telecom professionals say "T-1," they probably mean one of the following four definitions:

- *Access T-1:* A connection from the customer's premise to a long-distance carrier that provides access to the public-switched network. Access T-1s are generally provided by the local telephone company, but it is common for customers to receive a bill directly from the long-distance carrier (see Figure 15.5).

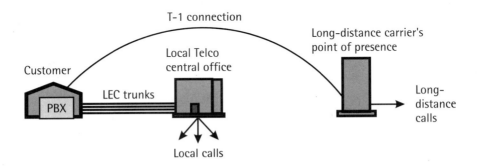

Figure 15.5 Customers use access T-1s to connect to their long-distance carriers.

- *Internet T-1:* A connection to an ISP. Medium to large-size companies use Internet T-1s for e-mail, Internet access, and Web hosting (see Figure 15.6).

- *Point-to-point T-1:* A dedicated private line between two customer sites. Customers use these fixed connections for data and voice traffic between locations. If voice traffic is carried on the T-1, this is called a *tie line*, because the T-1 ties the two locations together (see Figure 15.7).

- *T-1 Bandwidth:* Equals 1.544 Mbps, the capacity needed to carry 24 simultaneous phone calls.

T-3 service

T-3 service has 44.736 Mbps of bandwidth available. T-3s are normally only used by corporations and large universities. A T-3 has the capacity of 28 T-1s and can carry 672 simultaneous voice conversations. One T-3 usually costs the same as 8 to 10 T-1s.

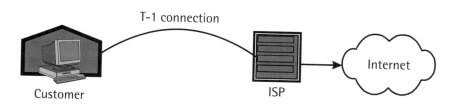

Figure 15.6 Customers use T-1s to connect to their ISP.

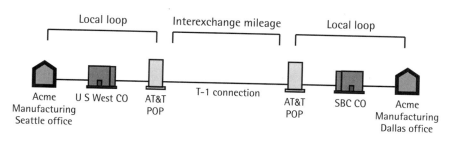

Figure 15.7 Customers use point-to-point T-1s to share data and voice traffic between locations.

All of these examples use dedicated private line technology. Circuit switching and packet switching are different technologies, but they are still measured by bandwidth. These technologies are explained in Chapter 16.

Sample bill

Figure 15.8 is a sample bill for a fictional dedicated private line between two customer locations in Seattle, Washington, and Dallas, Texas. Data

Figure 15.8(a) Sample data circuit bill: page 1.

Telephone Company B

Interstate Dedicated Private Line Service

Page Number: 2

ACME MANUFACTURING

Billing Number:	X5 X55555 01 001
Account Number:	8001-555-5555
Invoice Number:	555555 5555
Invoice Date:	07-01-2001
For billing inquiries:	1-800-555-5555

Description	Monthly Charges	Prorated Charges	One-Time Charges	Taxes and Surcharges	Total
Circuit Charges					
T1.5 MBPS SERVICE					
DHEC 555555 TCB #	$ 3,543.12	$ .00	$ .00		
Promotional Savings:	$ .00	$ .00	$ .00		
Discount Plan Savings:	$ 985.57 CR	$ .00	$ .00		
Net Charge:	$ 2,557.55	$ .00	$ .00	$ .00	$ 2,557.55
Total Circuit Charges:	$ 2,557.55	$ .00	$ .00	$ .00	$ 2,557.55
Total This Account:	$2,557.55			$.00	$2,557.55

Description	Total Promotional Savings	Total Discount Plan Savings	Total Interruptions	Total Service Assurance Warranty
Account Totals Reflect the Following				
T1.5 MBPS SERVICE	$ .00	$ 985.57 CR	$ .00	$ .00

Activity Occurred This Billing Period

Figure 15.8(b) Sample data circuit bill: page 2.

networking bills give little detail. Because the charges are fixed each month, the carriers must figure that the customers do not want much detail.

Figure 15.8(a), the first page of the bill, shows the customer has monthly charges of $2,557.55. Under the heading "Balance Brought Forward," we see that last month's charge was the same amount. A bill auditor notices these details because it confirms that this is a fixed monthly cost. The bill does not fluctuate because there is no charge for voice or data messages transmitted. This section of the bill also reveals that the customer still has not paid last month's bill. Bill auditors take note of this because carriers are less amicable with their deadbeat customers.

Figure 15.8(b), the bill's second page, reveals that it is T-1 service. The bill does not indicate whether or not this is an access T-1, an Internet T-1, or a point-to-point T-1. A seasoned bill auditor knows the monthly pricing for access and Internet T-1s is about $500 to $1,000, so this must be a point-to-point T-1.

The heading "Discount Plan Savings" indicates that the customer has signed a term contract and is getting a discount. An auditor would double-check the discount amount and find the customer receives a discount of 38.5%. The half-percent is suspicious; carriers usually give their discounts in whole numbers. It is possible that the customer is only receiving a partial discount.

To completely audit this bill, more information is needed. It is especially important to verify the addresses of the two ends of the circuit. Our hunch is that this is Acme Manufacturing's point-to-point T-1 connecting the Dallas and Seattle offices. For more detail on this circuit, we should match this circuit to the company network diagram and we should contact Telephone Company B.

Miscellaneous cost management strategies

Data networking is highly complex, but invoices from carriers are surprisingly simple. The next few sections offer a few miscellaneous practical strategies that nontechnical people can use to minimize the expenses of a data network.

Avoid billing errors with centralized control

Most of this book focuses on how a company can manage the *external* relationships it has with its telecom suppliers. Many large companies wind up overpaying because the *internal* relationships are mismanaged.

Most large organizations have an IT department that manages their companywide computer network. But the telephone bills associated with the network are managed by a separate department—the telecom department or the accounts payable department. The people who plan and order the services are different from the people who manage the costs. Even though the ones paying the bills might not understand what they are paying for, they can still successfully manage these costs.

The following is an example of a company whose internal processes ended up raising its telecommunications expenses. An aircraft maintenance company in Ft. Lauderdale, Florida, processed the telephone bills for all of the company's locations. The company's Austin, Texas, location ordered a new T-1 and informed the telecom department in Ft. Lauderdale. The Austin office handled all the negotiations and coordinated the installation with AT&T. Once the T-1 was installed, the Austin office was no longer concerned about pricing issues. The manager in the Austin office had an "if it ain't broke, don't fix it" attitude.

The Ft. Lauderdale office began receiving invoices for the new T-1, but it could not determine if the charges were correct. The Austin office misplaced the copies of AT&T's proposals and contracts, so it was impossible to verify the pricing. The corporate telecom department felt the charges were too high, but the manager in Austin wanted to ignore the situation because he was losing face. In the end, internal politics prevented the telecom department from efficiently managing the T-1 billing, and the company ended up overpaying AT&T for the entire 3-year term. The company's upper management should have established some strict guidelines for negotiating, ordering, and verifying all telecom services.

Free e-mail

Prior to the widespread use of fax machines and the Internet, businesses subscribed to e-mail service provided by carriers such as AT&T. The e-mail messages were transmitted across the carrier's network. Pricing for the service consisted of a monthly fee and a usage charge based on the number of characters sent. The service was expensive, but it was quicker and less expensive than overnight mail. Most businesses have replaced this type of e-mail service with Internet-based e-mail. A small number of businesses still have active accounts with carriers and still pay the invoice each month, even though the service is not used. The customer should cancel the service with the carrier and try to negotiate a refund for the previous few months' service.

Medium and large businesses pay their ISP for e-mail accounts in addition to the charge for monthly access to the Internet. Many ISPs will give their customers e-mail. This expense can also be eliminated by using one of the numerous free e-mail services available such as hotmail.com.

Avoid fraudulent charges

One of the latest telecom scams is "cramming" bogus Internet charges on a customer's local telephone bill. Most businesses do not question these charges and the thieves make easy money each month. LECs allow other companies to add charges to the LEC bill because it earns a billing fee. The charges are listed with a legitimate sounding name such as "Web hosting" or "Internet," and may be as high as $100 per month. In some cases, the thieves copy elements of the company's true Web site and build a phony Web site. The bogus Web site should be canceled, and the fraudulent company should give a full refund of all past charges.

Use a contingency plan

One of the fundamental strategies for managing a mission-critical data network is to have a backup plan, normally called a contingency plan. If the primary carrier's data network fails, then the data traffic can be redirected to a secondary carrier's network. In addition to being a backup plan, a two-carrier contingency plan also has cost management benefits.

For example, a Boston brokerage firm has a dedicated T-1 connection to Wall Street provided by WorldCom. The brokerage firm also installed a 56-Kbps line with Sprint to be used in case of a WorldCom service outage. The data networks of the telephone carriers rarely fail, so the brokerage firm may never use the Sprint 56-Kbps line. Nonetheless, using two carriers can be a strong negotiating tool for the business. When the WorldCom contract expires, the brokerage house will be able to negotiate very aggressive pricing with WorldCom. WorldCom would rather trim its profits on the account rather than lose it entirely to Sprint. Most small- and medium-size businesses do not bother with contingency plans.

Chapters 16 through 18 delve into more detail about the carriers' service offerings. In addition to dedicated private line technology, circuit switching and packet switching technologies are explained. The chapters also offer numerous cost management strategies for data networking.

16

Dedicated private lines

The concepts introduced in Chapter 15 are a cursory introduction to data networking. Many other sources are available for the ambitious student for further research. Assuming the reader is now familiar with basic data networking concepts, this chapter introduces the specific data networking services currently offered by telecom carriers. Most importantly, this chapter offers proven advice on how to minimize data network telecom expenses.

Technology evolution

Computers are constantly evolving into faster, more powerful machines. Data networks connecting these computers must adjust and evolve accordingly. Not only are computers more powerful, but they also are much more readily available today. Telephone networks host an ever-increasing amount of data traffic, and networks are frequently bogged down by traffic jams. The engineers who designed the telephone networks never expected so many guests to show up at the party.

To add more capacity to telecom networks, research and development engineers have developed new ways to send data across telephone lines. The three distinct generations of data networking technology are dedicated private lines, circuit switching, and packet switching.

Dedicated private lines

Dedicated private lines work like the toy phone children make by connecting two tin cans with a string. The tin can phone can only be used to communicate between two points across the string, such as between two tree houses. Like the string, dedicated private lines connect two separate points. For this reason, dedicated private lines are often called point-to-point circuits. Other names are dedicated circuits and leased lines. A state university, for example, may use dedicated private lines to share data between the main campus and its various satellite campuses.

The original data networks consisted of dumb terminals connected to a mainframe computer across dedicated private lines (see Figure 16.1). Data processing required a giant mainframe computer. It was more cost effective to have one mainframe computer at a central location than to

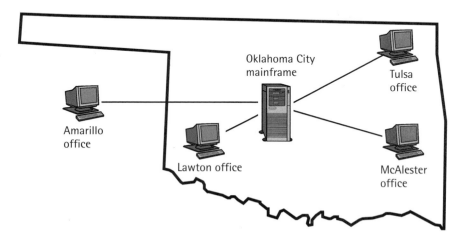

Figure 16.1 The original data network consisted of dumb terminals at remote sites connected to the mainframe computer at the home office.

have a computer at each remote site. The word "dedicated" means the phone company reserves the phone line from point A to point B just for you. The line is called private because no one else can use it.

The monthly cost of a dedicated private line is based on distance and bandwidth. The longer the distance, the greater the cost. The greater the bandwidth, the greater the cost. The bandwidth of most dedicated private lines is 56 Kbps, 256 Kbps, or 1.544 Mbps. Dedicated lines are expensive because the customer ties up the carrier's line all the time, preventing that line from being used by other customers.

The telephone companies appreciated the revenue from sales of dedicated private line service, but they feared that their networks would run out of room. As data networking began to be used by more and more businesses, the telephone companies could see that one day they would not have enough lines to meet demand. They had two choices: build more phone lines or figure out how to get more data across existing lines. They chose to do both. Engineers went back to the drawing board and came up with a brilliant solution: circuit switching.

Even though circuit switching (and eventually packet switching) is a more recent technology, dedicated private lines are still used in most data networks. Depending on the application, dedicated private lines require the use of the following types of data communication equipment: modem, channel service unit, data service unit, channel bank, or a PBX. A networking specialist normally helps a business decide what equipment is needed in a given application.

To lease or buy?

Data networks require the use of customer premise equipment (CPE). Some examples of data networking CPE are modems, channel service units, data service units, channel banks, network adapters, routers, and switches. Some businesses choose to purchase the equipment at the time of installation, while other companies prefer to lease, or rent, the equipment. The lease payment is normally included as a line item on the carrier's monthly bill.

Once a data network is up and running, most businesses do not give much attention to the invoices. Consequently, a business may end up leasing or renting a piece of equipment indefinitely. The business may be able to save money by purchasing the equipment outright, rather than making monthly payments.

The local loop

If the two endpoints of a dedicated private line are in the LATA, then the LEC will provide the service. For example, a hospital in Salt Lake City, Utah, has a 56-Kbps dedicated private line to a pediatrician's office across town (see Figure 16.2). Because the private line starts and ends within the same LATA, U S West provides the service. The same hospital may also have a 56-Kbps dedicated private line to a laboratory in Las Vegas, Nevada. Because this private line crosses LATA boundaries, an IXC, such as Sprint, provides the service.

This interlata private line consists of three parts: the local loop in Salt Lake City, the interexchange mileage, and the local loop in Las Vegas. The local loop in Salt Lake City is the connection from the customer's premise to Sprint's nearest central office in the same city. The IXC mileage is the middle portion of the line and runs between Sprint's central office in Salt Lake City and its central office in Las Vegas. At the other end of the line, in Las Vegas, the local loop is the connection from Sprint's central office to the customer's premise.

The two local loops are actually services provided by the LEC in each city, but all three charges will appear on one bill from Sprint. Sprint will procure these services for the customer from the LEC. Sprint will bill the customer for all three portions of the circuit. For the two local loops, Sprint will pay wholesale prices to U S West and Nevada Bell.

Local loop pricing is based on bandwidth and the distance to the central office. The further out the business is, the greater the cost. The bill for

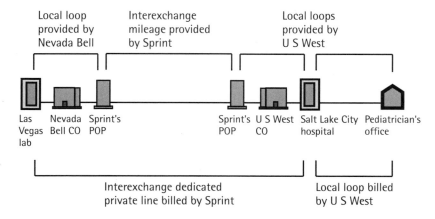

Figure 16.2 Local loops.

the middle portion of the private line is based on mileage and bandwidth. A good rule of thumb is that dedicated private lines generally cost $1 for every 1 or 2 miles. The following sections offer some basic strategies for saving money on dedicated private lines.

Check for missing discounts and promotions

When negotiating contracts for data services, telephone companies send out their best field sales representatives. Managers often get involved to ensure that the deal is closed. After the contracts are inked, the highly-polished, high-powered account team turns the service orders over to a data entry clerk, who is probably required to enter orders and answer customer service calls at the same time. The new service is installed, and the new account will be established with no glitches, but the special discounts and promotions frequently "fall off" the account. The end result is that the customer ends up paying the nondiscounted tariff list price for the service.

Most large businesses manage their telecommunications centrally. For example, a large Midwestern printing company with many offices around the country had complex data services with MCI. The telecom department at the corporate office negotiated specific discounts for MCI's data services. The pricing structure was complex and offered different discount percentages on different services. For example, access T-1s received a 10% discount, frame relay permanent virtual circuits (PVC) received a 15% discount, and interstate dedicated private lines received a 20% discount.

After the contracts were signed, it took MCI many months before the discounts appeared on the bills. Some locations never saw any discounts, or the discounts were applied at incorrect percentages. Access T-1s that should have had a 10% discount instead received only 5%. On some services, the discounts fluctuated each month. The bills for the Tampa, Florida, office may have been accurate in July, but in August, the discounts had vanished. In September, the discounts were back, but at incorrect percentages.

The situation continued for many months. Every month, the corporate telecom department spent days auditing the bills and weeks negotiating invoice credits with MCI. The root of the problem was never pinpointed but it probably had something to do with having such a large, multilocational account processed by a "legacy" billing system. Antiquated billing computers (called "legacy" because they are so old) are still used by most of the large carriers, because they are hard to replace when they are in use every month.

After a few billing cycles, the MCI account team was very discouraged by its inability to correct the problem, and the company was only responsive when the customer hinted at changing to another carrier.

Customers should scrutinize their bills for the first few months to ensure that all discounts are in place. If discounts and promotions are missing, the carrier should correct the errors and issue a refund of past overcharges. If the situation continues, the customer may be justified in switching carriers. Customers must carefully review their carrier service agreement to see if they are justified in canceling the contract and moving to another carrier.

Correct mileage errors

Dedicated private line pricing is based on bandwidth and mileage. The further the distance between the two points, the greater the cost. A T-1 between Baltimore, Maryland, and Baton Rouge, Louisiana, costs more than a T-1 between Baltimore and Buffalo, New York. The mileage is based on the exact distance "as the crow flies" between the two central offices at the endpoints of the circuit. Customers are sometimes misbilled because of incorrect mileage calculations.

In some cases, the carrier's mileage calculations are incorrect. A common sense review of the phone bill and CSR may expose an erroneous calculation. Most carrier account managers will take the time to help you double-check your mileage charges. If an error is found, the carrier will correct the pricing and refund the overcharges.

Disconnect dead circuits

When a telecom service is no longer used, the customer should immediately cancel the service. In many organizations, the telecom department quits using a service and forgets to tell the accounts payable department. Consequently, the phone bills continue to be paid every month. Many customers wrongly assume that no longer using a service tells the carrier, "We're done using this service, please quit billing us."

Carriers never close accounts unless a customer tells them to. If they did, they would risk canceling a customer's legitimate service. Because data accounts have flat-rate pricing, inactive accounts still generate the full amount of revenue for carriers. They make sure the customer has canceled the service prior to forfeiting this revenue.

For example, a New York-based travel agency paid for services it no longer used. Through a partnership with a nationwide travel company based in Philadelphia, the New York travel agent sold tickets for cruises, hotel rooms, and vacation packages. Each of its five offices in New York was equipped with a computer terminal that connected directly to the nationwide travel company's reservation office across dedicated private lines. The nationwide company supplied a special computer terminal, and the New York agency was responsible for paying for the dedicated private lines each month.

After a few years, the two companies decided to end the partnership. The nationwide travel company picked up the computers, and the New York travel agency figured the project was closed out. The agency's accounts payable clerk continued to pay for the service for the next 2 years. The company never canceled the dedicated private lines with its carrier.

Years later, an outside auditor noticed the data lines terminating in Philadelphia. Once he found out the two companies were no longer business partners, the services were canceled. Correcting this error created thousands of dollars in monthly savings for the company. Even though the problem was the customer's fault, the carrier still issued a significant "good faith" refund to the customer.

Cancel dangling circuits

When a customer cancels a point-to-point data circuit, carriers sometimes fail to disconnect the circuit at both locations. This results in being billed for a dangling circuit, which is like the tin can phone with only one can. Dangling circuits are most prone to show up when a private line is furnished and billed by two separate carriers.

Phone bills for dedicated private lines usually do not give enough detail to detect dangling circuits. You must call customer service and ask for the addresses of both circuit locations. You can also check the CSRs for the circuits. Local carriers and some long-distance carriers, such as AT&T, keep CSRs. The carrier's CSRs always specify the locations of the origination and termination points of a dedicated private line. If the CSR only lists one location, it is probably a dangling circuit that should be disconnected.

Network diagram audit

An organization's IT department usually maintains a network diagram. If a circuit is no longer used, the IT staff erases it from the network diagram. If the staff forgets to cancel the billing with the carrier, the organization may go on paying the bill for years. An effective way to locate unused data lines is to match up the company's internal network diagram with the actual phone bills. The diagram reflects exactly what data services are in use at each location, and the phone bills reflect the actual services that are currently being billed.

Consolidate circuits

When numerous data circuits terminate at the same customer location, it is possible to consolidate the smaller circuits into one large circuit, especially in the case of point-to-point circuits. As previously explained, the first leg of a dedicated private line is the local loop. In many cases, a business can consolidate its various local loops into one circuit with greater bandwidth.

For example, a manufacturer in Marietta, Georgia, was expanding into other states. When a new facility was opened, a new 56-Kbps dedicated private line was installed from the Georgia office to the new facility. The cost of each 56-Kbps line was based on the local loop charges in Marietta, the IEX mileage, and the local loop charges at the new location (see Figure 16.3). The Marietta local loop charge was $150 on each of the five lines. In total, the business paid $750 for 280 Kbps of bandwidth.

When the company had five dedicated 56-Kbps lines, the BellSouth account manager helped it cut its cost. The business consolidated all of the traffic from the five data lines onto a single T-1 data line that gave the business four times more bandwidth at a cost of only $350 per month (see Figure 16.4).

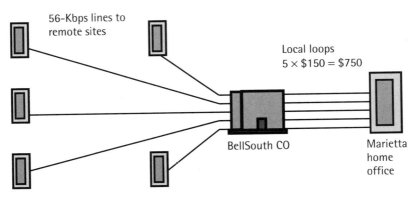

Figure 16.3 A company in Marietta, Georgia, has five data lines connecting to remote sites in other states.

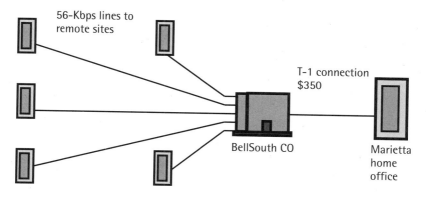

Figure 16.4 Five local loops are replaced with a single T-1 connection.

In another example, a pharmaceutical research facility in Wisconsin paid more than $10,000 for a dozen T-1s used for voice and data. The telecom manager cut the cost in half by consolidating all of the T-1s onto a single T-3.

Volume and term discounts

Dedicated line charges are nonfluctuating charges, so volume discounts usually do not apply. Data services are more subject to term discounts. The simplest way to cut the cost of data services is to sign a term agreement with the carrier. Unlike other telecom services, data term agreements can be as long as 7 years. Table 16.1 shows how data circuit discounts are often structured.

Table 16.1 Typical Monthly Pricing for Data Circuits

	TERM CONTRACT DISCOUNT							
	No Term	1-year	2-year	3-year	4-year	5-year	6-year	7-year
	0%	5%	7%	10%	12%	15%	20%	25%
56 Kbps	$300	$285	$279	$270	$264	$255	$240	$225
256-Kbps fractional T-1	$500	$475	$465	$450	$440	$425	$400	$375
T-1	$1,000	$950	$930	$900	$880	$850	$800	$750

Update pricing

The tariffs filed with state and federal regulatory bodies describe carriers' service offerings and detail pricing information. From time to time, carriers update their tariffs. Prices usually go up, but sometimes they go down. The phone companies also might add a short-term promotion, to stimulate sales activity.

Once or twice a year, you should contact your carrier and find out the latest pricing for the services they currently use. If the new pricing is lower than your current pricing, ask the carrier to upgrade you to the latest offering. Carriers always say "this pricing is only available for new customers, not existing customers." Sometimes, however, a sympathetic account manager will implement the new low pricing anyway. It helps if the customer has some leverage in the negotiations.

LATTIS.PRO

Some consultants use LATTIS.PRO to check circuit pricing. The software instantly tells you prices for circuits throughout North America. The prices will be broken down according to each circuit element. LATTIS, which is a fee-based service, is located on the Internet at www.triquad.com/lattis.html.

Add promotions to the account

Consider the following example: A Sacramento, California, publisher had a dedicated T-1 line connecting to a marketing office in Phoenix, Arizona. The T-1 was provided by Qwest and cost $1,365 each month. The publisher learned from a friend that Qwest was waiving local loop charges in Sacramento for all new orders for dedicated T-1s. Armed with this information, the publisher approached Qwest and requested the promotion.

The publisher's long-distance contract with Qwest was about to expire. The account executive did not want to lose the account, so he added the promotion to the publisher's T-1 bill. Waiving the local loop charges saved the business $185 per month.

17

Circuit switching

The original telephone service consisted of two phones in two separate locations connected by a single line. To call multiple locations required multiple phone lines. Cities across America began to be covered with a network of unsightly telephone wires. After the switchboard was invented, a person could call any other phone using a single telephone line. Half a century later, the same situation occurred, except this time it was with data calls instead of voice calls.

With dedicated private lines, two remote computers connect over a distance using a fixed circuit. That circuit cannot be shared by anyone else. But the phone companies do not like their lines being tied up, so they invented circuit switching.

With circuit switching, the caller (normally a computer) dials the other caller and the two have exclusive use of the phone line until they decide to end the communication. Once the communication is finished, the connection ends and the line is available for another caller. Circuit switching works just like a regular voice phone call between two people; they call, they chat, they hang up.

At the beginning of each circuit switched call, the network determines the route of the call. That path, or circuit, is open for the duration of the call. On the next call, the network may choose an alternate path (see Figure 17.1). With packet switching, the network establishes a permanent route for the call. On each call, the data travels across the same path in the network. In frame-relay networks, this is called a PVC.

ISDN

The most common type of circuit switching used in business today is an ISDN. "Integrated services" means that a user can send voice, data, and video across the network at the same time. "Digital" refers to the fact that the lines provided by the telephone company transmit digital, not analog, signals. Digital is cleaner and faster than analog. Computers using a normal analog phone line use modems to convert between the computer's digital signal and the phone line's analog capabilities. ISDN, being digital, requires no modem but does require the use of a *network adapter*. It looks just like an external modem but costs three times as much.

ISDN comes in two sizes: large and really large. Basic rate interface (BRI) is normally used by a single person, while dozens of employees may use one primary rate interface (PRI) connection for voice and data traffic. BRI is used for applications such as telecommuting, Internet access for a single user, and, occasionally, videoconferencing. PRI has the same bandwidth as a T-1 and is used for similar purposes: to connect multiple users to the Internet or to carry voice calls from a PBX to a carrier.

BRI

ISDN technicians say BRI is "2B + D," meaning the service has two separate bearer channels for transmitting information and one data channel used

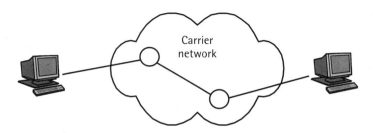

Figure 17.1 Circuit switching.

behind the scenes for signaling. The bearer channels have 64-Kbps bandwidth each, and the data channel has 16-Kbps bandwidth.

PRI

PRI is "23B + D." There are 23 bearer channels and 1 data signaling channel. These 24 channels are the standard used in the United States and in Japan. Elsewhere in the world, PRI ISDN uses 30 channels. One of the main advantages of the D channel used in PRI ISDN is that an inbound caller's phone number can be passed along to the PBX, which is very helpful for call centers.

ISDN pricing

ISDN installation costs vary widely from carrier to carrier and can sometimes be waived. After the customer has purchased the equipment and paid for the installation, the following monthly recurring charges apply with both types of ISDN: line charges, local calls (if measured service), and long-distance toll calling.

Monthly line charges for BRI ISDN are between $50 and $150. PRI ISDN line charges are normally between $300 and $1,000. The charges for local calling and long-distance calling are about the same as standard voice telephone call rates.

ISDN began to be widely offered by the telephone companies in the late 1980s. Although the service promised many improvements over previous technologies, customers were not eager to subscribe. Some blamed poor marketing by the telephone companies, but the real reason is probably because the initial service offerings were too confusing to business people. The phone companies required prospective ISDN customers to be too involved in minor technical decisions. The whole process was a turnoff to business people, and ISDN got a bad name. Some said ISDN really stood for "innovations subscribers don't need." Over the past few years, the telephone companies have made new activations less painful, and the demand for fast connections to the Internet has breathed new life into ISDN service.

ISDN loose traffic

ISDN is a measured service, and users pay for each minute of calling in addition to a flat monthly fee. If a new ISDN customer does not inform the long distance carrier of the new ISDN line, then the calls will be billed on the LEC's ISDN bill. The calls will be billed at casual rates, which may be

$0.35 to $3 per minute. Customers whose ISDN calls are being billed at casual rates on their LEC bill should inform their long-distance carrier, which can move the ISDN loose traffic to the long-distance bill. This should result in much lower pricing.

Although circuit switching was an improvement over dedicated private lines, phone company upper management still was not satisfied and continued to worry about traffic jams on their telephone network. The biggest problem with circuit switching, from a telephone company perspective, is that, similar to regular voice calls, the phone company circuit is tied up for the entire length of the phone call. During a regular voice call, if one person sets the phone down to check the roast in the oven or answer the doorbell, the phone line is still tied up. Even though the line is quiet and no data is being passed across it, the line is still tied up and no one else can use it.

When thousands of businesses across the country make simultaneous circuit switching calls, there may eventually be a shortage of capacity. Once again, phone companies sent their engineers back to the drawing board with a similar charge "to figure out a way to send even more data across the existing phone network." The end result was packet switching.

18

Packet switching

Dedicated private lines tie up an entire phone line for the entire month. Circuit switching also ties up an entire phone line, but only for the duration of the call. Packet switching is much more efficient. Packet switching only ties up part of the phone line for the duration of the call. Like circuit switching, if the computers are not sending or receiving data, they stay off the phone lines, keeping them free for other users. But with packet switching technology, even the quiet times within one call will be filled with packets of data from someone else's call.

The data is broken into packets that are sent from point A to point B. The packets may travel across different paths within the telephone company's network. At the terminating point, all the packets are reassembled in the correct order.

What is a packet?

A packet is group of bytes of information that are processed independently across packet-switching networks (see Figure 18.1). Each packet has three parts: header, data, and footer. The header and footer are like a train with

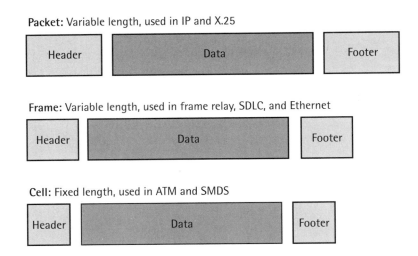

Figure 18.1 Packets, frames, and cells.

an engine in the front and a caboose in the back. They contain important information about the packet, such as the sender's address, the destination address, the size of the packet, and the type of data contained in the packet, such as voice, data, or video. Voice and video packets are given priority in packet-switching networks, just like passenger trains are given priority on a railroad. Voice and video require a steady stream for the transmission to be smooth.

One reason packet switching is fast is because it does not correct transmission errors within the network. Instead, the data is sent again. The receiving computer tells the sending computer "I didn't get all the packet ... please resend the missing ones." This is very effective for data communication where slight delays can be tolerated, but most businesses still refuse to move their voice traffic across a packet-switching network, because of the slight delay. Few companies want to sound like the 1980s computerman Max Headroom when they talk to their customers.

Prior to being transmitted across a computer network, a given computer file, such as a simple word processing document will be broken into packets of data. For example, imagine that a man named Sam in Seattle decides to write a love letter to Louise in Long Island. Sam plans to transmit the letter to Louise across a packet-switched network.

Using Microsoft Word, Sam writes a brief letter to Louise expressing his deep love for her. Sam is not a man of many words; his letter contains only 100 words. The size of the computer file containing the letter is only 20 Kb. Sam's company uses a packet-switching service called ATM. The ATM service breaks Sam's letter into 377 individual packets and transmits each one to the Long Island office. The whole process takes less than 1 second. In Long Island, the packets are reassembled in order, and the computer file is now accessible. Tears run down Louise face as she reads Sam's love letter.

Frame relay

The original WANs consisted of multiple remote locations connected together. Each site may have a single computer or an entire LAN. The LANs connect to each other with dedicated lines provided by a telephone company. But the dedicated lines are expensive for customers, and they eat up too much of the carrier's network capacity. Frame relay service is a solution to both of these problems.

Frame relay service uses variable-sized packets of data called frames. Unlike X.25, an earlier packet-switching service, frame relay service is a fast-packet technology. It discards erroneous packets, instead of correcting them. Error correction is performed at the end-points only, and not along the way, which results in a speedier transmission. If the receiving computer detects errors, it asks the sending computer to retransmit the data. Because errors are few with digital communication, this feature makes frame relay very quick. Unlike dedicated lines, frame relay is not a full-time connection. This is not a problem, because most businesses transmit data intermittently.

The PVC

Frame relay service providers set up a PVC between two customer sites that acts like a dedicated line. The customer chooses each PVC's bandwidth. For example, a Louisville furniture manufacturer uses a 56-Kbps PVC to a remote facility in rural Kentucky but has a 256-Kbps PVC to the sales office in Atlanta, Georgia. More bandwidth is needed to the sales office, because more data is shared with this site. If the company used dedicated lines instead of frame relay, the cost would be almost double. Frame relay gives the best of both worlds: lots of bandwidth and low cost.

The CIR

The frame relay customer chooses the bandwidth of each PVC. As with other data services, the higher the bandwidth, the higher the cost. The 256-Kbps PVC costs more than the 56-Kbps PVC. The bandwidth is called the committed information rate (CIR), which is simply the rate of information that the phone company commits to always have available for you. The frame relay CIR is, therefore, the *minimum* speed limit. On the other hand, a 56-Kbps dedicated private line can transmit data no faster than 56 Kbps. The bandwidth of a dedicated line is, therefore, the maximum speed.

Frame relay is a "bursty" service. The furniture company could potentially transmit data at T-1 speeds across a 56-Kbps PVC if the phone company network has some spare bandwidth. Qwest boasts that its network has so much capacity that customers can save money by specifying "zero CIR" and still transmit data at T-1 speeds.

If frame relay service is used all within one LATA, then the service will be provided by the LEC in the area. For example, a Seattle hospital with numerous clinics in the same metropolitan area would purchase frame relay services from either its local carrier U S West or a competitive LEC that operates in the area. If the data network crosses LATA boundaries, a long-distance carrier, such as AT&T, will provide the service. However, the customer will still pay a local loop charge. The local loop is a dedicated private line from its facility to the long-distance carriers nearest frame relay-equipped central office. The local loop is provided by the LEC but will be billed on the frame relay carrier's bill.

To install frame relay service, the customer must buy routers to be used at each location. Each carrier charges an installation fee, and the network technicians who program the router may charge additional fees. A business' monthly frame relay pricing is based on PVCs, CIR, local loop charges, and any contractual discounts. Frame relay is a measured service—carriers show the usage on each invoice but most do not charge for it.

Replacing dedicated lines with frame relay

Consider the following example: A chain of tire stores in Ohio has four locations. Each location connects to the other three with dedicated T-1 lines. This "fully meshed" network requires a total of six T-1s. The charge for each T-1 includes the local loop at the starting point, the interexchange carrier mileage, and the local loop on the terminating end of the circuit. The network is illustrated in Figure 18.2.

The company decided to replace its network of T-1 lines with frame relay service. Using frame relay, each location only requires one local loop connection to the frame relay provider's central office. The data is then transmitted across the carrier's network, which is usually called a cloud. To convert from dedicated lines to frame relay, the company had to purchase routing equipment and pay installation fees. The monthly charges are based on the local loop charges, the PVCs, and the CIR chosen by the customer. Figure 18.3 shows the change from dedicated lines to frame relay. The customer's monthly cost dropped from $6,000 to $3,000.

Voice and data convergence

One of the recent data networking trends is called *convergence*. Convergence means that different signals such as voice, video, and data are transmitted over a single medium. On the consumer level, Web TV is an example of convergence; cable TV and the Internet are provided across a

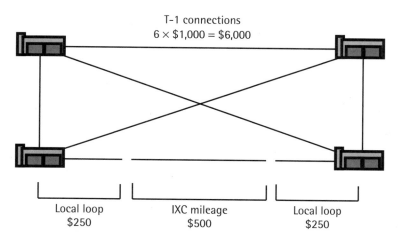

Figure 18.2 Meshed network.

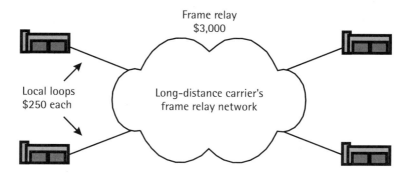

Figure 18.3   Frame relay network.

single cable. At the business level, many companies are migrating their voice long-distance traffic across their data networks.

Businesses have traditionally carried their voice and data traffic over separate networks. The phone companies assigned separate account managers to handle a business' voice and data needs. The services had separate contracts and were billed separately. As part of the recent streamlining effort of the phone companies, voice and data services now share the same account manager, the same contract, and the same invoice. But the greater change is that voice and data now ride the same phone line.

Voice over frame relay

The main benefit of voice over frame relay (VoFR) is that long-distance calls are free. VoFR is normally only used to carry intracompany long-distance traffic or international long-distance traffic. The *latency* of packet-switching technology affects the call quality (i.e., conversations may be choppy).

Latency is a term that describes the transmission delay due to the speed of the media and the processing time of the network equipment, such as routers. Each stage in the network may only add milliseconds of delay, but the combined latency may be enough to distort the sound of the phone call. VoFR is, therefore, rarely used for "front office" applications. However, businesses with lots of intracompany long-distance calling can significantly reduce their long-distance billing. By moving their long distance across their frame relay network, they will eliminate the long-distance cost altogether. If they do not mind slightly compromising call quality, thousands of dollars can be saved.

Voice over Internet

Voice over Internet (VoIP) is the same concept as VoFR, except the voice calls are converted to data packets and sent along a network that uses Internet Protocol (IP). The Internet, or a private WAN using IP, are both examples of IP networks. The same latency problems previously described with VoFR apply with VoIP.

On a much smaller scale, cost savings can be achieved by using one of the Internet's free long-distance Web sites such as http://www.net2phone.com or http://www.dialpad.com. These services allow a person to make free long-distance calls over the Internet. Most of them limit the destination of the call to the United States, but a few of these services have roots in the Far East and may include Korea or Taiwan as approved calling destinations. Internet phone calls often experience a lot of noise, similar to a shortwave radio conversation, or international calls 5 or 10 years ago. The call quality is poor, but you cannot beat the price.

ATM

ATM is a high-speed packet-switching telecommunications service. ATM is typically used only by very large businesses such as Fortune 100 companies, major universities, and telephone companies. Telephone companies use ATM technology in the "backbone" of their networks. A voice phone call from New York to Tokyo will probably be converted to ATM packets as the data travels along an undersea phone line lying at the bottom of the ocean.

ATM is a high-cost service, designed for high-volume users, and will therefore not be used by most businesses. According to the Vertical Systems Group, only slightly more than 35,000 enterprises worldwide are currently using ATM services, while frame relay has more than 1.2 million subscribers.

Why is ATM so fast?

ATM carries voice, video, and data at speeds up to 622 Mbps. Such a high speed is due to three factors: asynchronous switching, cell length, and the use of hardware in switching. "Asynchronous" means the service transfers different data at different times and can process multiple jobs simultaneously.

ATM's fixed-length packets, called *cells*, make ATM more efficient than other technologies, such as frame relay. The size of each frame relay packet must be processed, while ATM networks waste no time figuring out how large or small a packet is. ATM networks expect each packet to be 53-bytes long, and they rapidly move these packets up and down the network.

Another advantage of ATM over frame relay is that the switching is controlled in the network hardware, instead of the software. These three advantages make ATM a very fast data networking technology.

Although ATM may be a fast technology, it is also a costly one. Installing an ATM network is very expensive, and the monthly charges not only include fixed charges for the network, but also usage on an ATM network. ATM is only a cost-effective technology for extremely large businesses.

19

New data networking services

New technologies, such as the Internet and video-on-demand, have caused a strong hunger in the marketplace for more bandwidth. Everybody wants to send more data at faster speeds. Many large companies pay for a T-1 connection to the Internet, but smaller businesses do not spend their money as freely. Most residences and many small businesses are far away from the telephone company central office, so it is too costly for carriers to offer advanced data services at an affordable price. The little guys have been left out—until recently.

DSL and cable modems: High speed and low cost

During the last few years, the phone companies have looked for new ways to offer high bandwidth services to small businesses and consumers. The two most prominent technologies that have recently stormed the market are digital subscriber line (DSL) service and cable modems. Both services offer bandwidth up to 1.544 Mbps for less than $100 per month. That means a consumer can get T-1 bandwidth without paying $1,000 a month for it.

DSL service is being widely adopted by both small businesses and consumers. Cable modems have been most attractive to consumers, probably because most homes are already wired for cable TV service.

Both DSL and cable modem service are dedicated connections. The line is always available for Internet use. Cable modems do not have to "dial-up" the specific ISP; they are always connected to the ISP. In addition to unlimited Internet access, most DSL and cable modem service providers give their customers e-mail accounts and Web site hosting as part of the monthly service.

According to Computer Economics, cable modem subscribers in the United States will increase from 5.7 million in 2000 to 27.6 million in 2005. DSL subscribers are expected to increase from 2.4 million in 2000 to 13.8 million in 2005.

DSL

As previously explained, the "last mile" of copper wiring from the telephone company's central office to a business or residence has limited bandwidth capacity. End users demand lots of bandwidth, and phone companies want to earn revenue from this opportunity. DSL technology is a recent development that should satisfy both consumers and phone companies.

DSL deployment began in 1998. Since then, both computer manufacturers and telephone companies have not yet ironed out a single standard for DSL service. Consequently, numerous flavors of DSL are being offered today. The whole family of DSL services is often referred to as xDSL, with the "x" representing any number of other letters such as ADSL, CDSL, UDSL, VDSL, G.Lite, DSLLite, and freeDSL.

DSL offers a lot of bandwidth to small companies and consumers at phenomenally low rates. A typical DSL customer can receive 1.544-Mbps bandwidth across an ordinary copper telephone line for around $50 per month. That amount of bandwidth has previously only been available to businesses that paid as much as $1,500 per month for T-1 service.

DSL technology sends a digital signal across a traditional twisted-pair copper telephone line. Because the signal is never converted to analog, greater bandwidth is available. Like ISDN, DSL service can carry both voice and data simultaneously. A person can surf the Internet and talk on the phone at the same time. DSL uses a dedicated connection to the Internet.

Upstream and downstream

DSL has separate transmission rates for "upstream" and "downstream" data. A person surfing the Internet will receive large amounts of data "downstream" from the Web site because of the numerous graphics and files. The amount of data sent "upstream" is minimal, because the Internet surfer is only sending mouse clicks or occasional keystrokes. DSL service typically offers upstream rates of 128 Kbps, and downstream rates of 1.544 Mbps. Figure 19.1 illustrates DSL's differing rates for upstream and downstream traffic.

A DSL customer that wants to use his telephone and computer on the same line must have the signal separated so that the bandwidth can accommodate the phone's analog signal and the computer's digital signal.

DSL is available across the United States in most large and mid-size cities. Flashcom, one of the largest DSL providers, offers service for $49 per month. This includes Internet access, and if the customer signs a 24-month term agreement, the equipment is free and installation fees are waived. In Missouri, SBC provides DSL service for $39 a month, or $49 a month with Internet access included. A 12-month term agreement is required with this pricing, and the customer must purchase a "DSL modem" for $198. DSL is a flat-rate service; there are no monthly charges for usage.

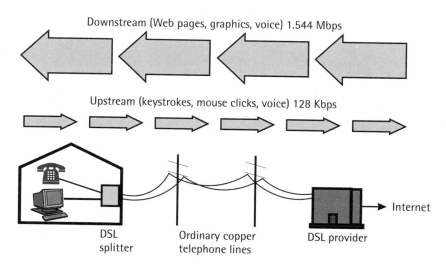

Figure 19.1   DSL service uses standard copper telephone wires but can deliver T-1 bandwidth downstream from the carrier.

Save money with DSL

Besides providing a large amount of bandwidth for Internet usage, DSL can also be used to lower existing telecom costs. A small independent insurance agency in Baltimore, for example, recently signed up for DSL service with a national DSL provider. The service provided a 512-Kbps dedicated connection to the Internet, e-mail, and Web site hosting, and the business could still make analog phone calls over the same line. The business previously paid more than $300 per month for all of these services. See Table 19.1 for a cost comparison of this change.

Cable modems

Cable modems are designed for high-speed Internet connections. Like DSL service, cable modems provide Internet access at different speeds downstream and upstream. The actual bandwidth for Internet connections over a cable line is 27 Mbps downstream and 2.5 Mbps upstream, but the total rate experienced by the end user is normally 1.544 Mbps. That is not too bad, though, considering that today many businesses pay more than $1,000 a month for this amount of bandwidth.

A new cable modem customer needs a service provider and the cable modem itself. Most service providers allow the customer to rent a cable modem; otherwise, the customer must pay around $300 for the device. With Time Warner's Road Runner service, the subscriber pays a one-time installation fee of $100 and a monthly fee of $40. The monthly fee includes rental of the cable modem and unlimited Internet access. Cable modem users do not pay hourly fees for Internet use.

Table 19.1   Saving Money With DSL

Current Services		Digital Subscriber Line
Phone line	$35	$49
512 Kbps fractional T-1 to the Internet	$235	
Web-site hosting	$50	
E-mail	$25	
Total:	$345 Total:	$49
Monthly difference:		$296
Annual difference:		$3,552

SONET and DWDM

Chapters 16 through 18 have explained the three basic types of data networking in use today: dedicated private lines, circuit switching, and packet switching. Specific services that use these technologies include ISDN, frame relay, ATM, and DSL. These services are commonly used by businesses.

Two other data networking technologies bear mentioning here: synchronous optical network (SONET) and dense wave division multiplexing (DWDM). Both SONET and DWDM are technologies used for transmitting data across fiber-optic lines. SONET and DWDM are used within carrier networks and rarely in a customer's network. A single SONET connection is capable of simultaneously carrying 129,000 conversations. These carrier-based services are too high-end to be covered in a cost management book written for business professionals. They are only mentioned here to inform the reader because telephone companies often boast that their networks use these services.

20

Wireless telecommunications service

More than 100 million Americans use wireless phones. Almost every household has at least one cellular phone. Within the next 10 years, more than 75% of all Americans will own a wireless phone. Around the world, especially in developing countries, wireless phones are becoming more prevalent than landline phones. In China, for example, more than 100,000 people purchase wireless phones every day.

As in other rapid-growth industries, wireless carriers are often negligent toward their customers. When customers sign up for new wireless service, they usually waste money by buying the most sophisticated phone, and they pay too much by selecting inappropriate rate plans. Sooner or later, every wireless phone user is overcharged because they fail to proactively manage their account. Corporate wireless users consider cell phone bills trivial, and consumers either do not know how to decipher the bills or they would rather spend their time doing something else. The carriers also regularly change their offerings, so it is very difficult for a customer to stay on the most cost-effective plan.

Chapters 20 through 23 explain the fundamentals of the wireless industry and offer proven techniques for effectively controlling these expenses. First, let us examine the wireless telecom industry's background.

Mobile radio and mobile telephones

Two-way mobile radio communication was first used around the time of World War I. Mobile radio is still used today across the globe by taxicab companies, police departments, fire departments, trucking companies, and marine operations. Mobile radio users only communicate with each other; they cannot connect to the public-switched telephone network. Users take turns speaking because the system only allows for one-way communication. But the price of *wireless telephone* service is dropping and, in most cases, it is now cost effective for a company to replace its mobile radios with cellular phones.

Mobile telephones were developed because people wanted to call from their cars to normal landline phones. The first mobile phones were very expensive, and each city could only handle a small number of simultaneous phone calls. One of the first mobile phone systems, in St. Louis, could not even accommodate 100 simultaneous calls. Mobile telephones were given their own frequency band by the FCC, and this small portion of the airwaves would only accommodate a small number of users. Cellular telephone service changed all of this.

Cellular service

Cellular telephone service has more capacity than the previous mobile telephone service because it divides an area into hexagonal shaped *cells*. Each cell uses different frequencies from the adjacent cells. The limited number of frequencies could be recycled in other cells, which allowed more simultaneous calls. Cellular telephone service is an analog technology. Figure 20.1 illustrates the different cells in an area.

In the early 1980s, the FCC divided the country into 734 distinct cellular service markets. In each market, the FCC divided the available airwaves into an *A-block* and a *B-block*. Two cellular service providers would operate in each market. The A-side carrier was usually an independent company such as Cellular One, while the B-side carrier was operated by the local telephone company, such as BellSouth Mobility. Even today, there are only two providers of cellular service in each market. But in most markets, both the A-side and B-side carriers allow other companies to resell their service, which results in more vendor choices for the customer. In the Midwest, Ameritech Cellular resold GTE's service until it built its own digital

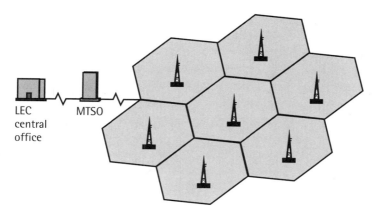

Figure 20.1 Cell map.

network. Most consumers are unaware that cellular companies may share the same network with their competitors.

For example, a construction supervisor who worked in a low valley was disappointed with his carrier, GTE. His work area was obscured by hills that prevented the cell phone from receiving the signal from the radio tower. He was, in effect, in a "dead zone," which is an area that cannot receive the signal from the tower due to an obstruction, such as a hill (see Figure 20.2). The construction supervisor switched to Ameritech Cellular

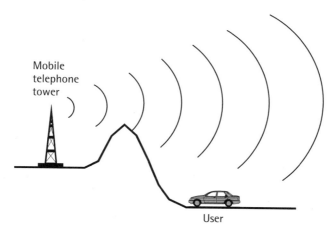

Figure 20.2 Dead zone.

but was disappointed because his phone still did not work in the valley. He switched companies, purchased a new phone, and changed phone numbers, but his coverage was exactly the same because the new carrier used the same network as the old carrier.

When traditional analog cellular service became popular, cellular providers needed to increase the capacity of their networks. Initially, they just divided the cells into smaller cells, which allowed them to carry more calls but also resulted in more dropped calls, especially near cell boundaries. Cellular customers can be fickle and will quickly change carriers if they have trouble making calls.

Customers were also disappointed with the insecurity of cellular networks. Using analog scanners, eavesdroppers can easily listen to a cellular phone conversation. The most famous case involving this was when an elderly couple listened in to Congressman Newt Gingrich's cellular calls.

Even more disappointing for the cellular customer is the practice of *cloning*, when perpetrators *clone* a cellular telephone by learning its electronic serial number (ESN). This ESN is then reprogrammed into another handset, and the perpetrator makes unlimited calls on this phone. No one will find out until the charges for these calls show up on the innocent customer's bill. The lack of security with analog cellular service drove many customers to switch to digital cellular service.

Digital cellular service

In the late 1990s, carriers began migrating their customers from analog service to digital cellular service. Digital cellular allows carriers to increase privacy, reduce cloning fraud, and increase capacity by 3 to 10 times. Digital cellular signals are multiplexed and scrambled, making it more difficult for eavesdroppers to listen in to a conversation. Digital service also includes more features such as caller ID, call waiting, repeat dialing, and call return. A carrier that upgrades from analog to digital technology can offer better service to more customers.

PCS

In the early 1990s, the FCC auctioned off six more bands of the airwaves to be used for personal communications service (PCS). The auction was controversial and the stakes were high. Many people complained that the government had no right to collect money by "selling" the airwaves. The bids at the auction were much higher than predicted and the government was surprised at how much money it raised. The auction of the A, B, and C blocks netted the U.S. Treasury $17.9 billion. In the Washington/Baltimore market alone, AT&T Wireless paid $211,771,000 just for the right to broadcast phone calls. Some of the entrepreneurs who got the bids eventually defaulted on the payments. Consequently, PCS service was not rolled out as quickly as planned.

PCS service uses digital technology and is rich with features. PCS also uses less complex pricing and billing schemes than traditional cellular service. Because of these advantages, wireless carriers and customers alike are making the shift from analog wireless services to PCS.

Low Earth-Orbiting Satellites

Low earth-orbiting satellites (LEOS) provide wireless telephone service for users across the globe. The handheld phone transmits and receives signals to and from geosynchronous satellites. Once enough satellites are launched into space, a caller should be able to make and receive calls anywhere across the globe. This revolutionary technology will allow people everywhere to make voice calls, transmit data, or connect to the Internet from anywhere. The concept is especially attractive in isolated places such as the Andes Mountains, where the telecommunications infrastructure is significantly underdeveloped.

LEOS technology has many glitches, however. During the recent war in Kosovo, telephone lines were severely damaged. News correspondents brought in satellite phones to call the outside world. In spite of the hefty $2,000 to $3,000 price tag, these phones rarely worked properly, especially when used indoors.

Probably the biggest problem facing rapid LEOS deployment is the high cost of building the network. In 1998, the Iridium Company, daughter company of Motorola, first offered global wireless telephone service. Iridium spent more than $5 billion building its 66-satellite constellation that would provide coverage throughout the world. But in its first 2 years, the company enrolled only 10,000 customers. A small customer base means small revenue, and in 2000, Iriduim filed for bankruptcy. What began as a dynamic cutting-edge high-tech company quickly died as the result of financial problems. Even its major supporter, Motorola, refused to bail the company out. Iridium is now back in business, but it mainly targets commercial users in remote parts of the world, such as off-shore locations.

Number portability

In the 1990s, the FCC mandated that the wireless phone industry must implement a number portability system. Historically speaking, cellular phone numbers have not been portable. If a user changes from Bell Atlantic Mobile to Sprint PCS, she has to get a new phone number. The old number is recycled by Bell Atlantic Mobile and Sprint PCS assigns the user a new number.

This process is so frustrating that most customers simply stay with the current carrier, rather than experience the "pain of change" to a new carrier. Getting a new phone number may require a business user to print new business cards and new letterhead, and the user must inform his clients of the new number. Few people are willing to go through this headache just to save a few dollars each month.

Third-generation technology

Telecom suppliers worry about the capacity of their networks. If the network is maxed out, they cannot enroll new customers, and existing customers switch to different carriers. New technologies are implemented to increase network capacity and allow carriers to offer more sophisticated services to their customers. The next significant wireless telecommunications technology is referred to as "the third generation."

Three 3G technologies have been developed, but none has yet been deployed. The three technologies are WCDMA, CDMA2000, and TD-SCDMA, developed by Americans, Europeans, and Chinese, respectively. Governments, telecom manufacturers, carriers, and other power brokers are still deciding which technology to deploy. 3G is often in the news because of the high-level negotiations deciding the future of the wireless industry, but until it is deployed, 3G is not very relevant for the end user.

WAP

WAP allows mobile telephone users to send and receive electronic data. WAP has been heavily marketed by carriers, but the service has had a rocky start. The service had many technical glitches, and the technology's early adopters lost their enthusiasm. WAP's biggest drawback is that it is not user friendly. Data input is inconvenient, and most customers cannot tolerate the tiny display screen.

Sample mobile telephone bill

Figure 20.3 is an example of a typical wireless telephone bill. The fictional user has a Telephone Company D phone in the St. Louis market. Mobile phone bills have three sections: cover page with payment coupon, account summary, and the call detail. In this particular bill, the customer pays $50 per month for access and gets 500 minutes of free airtime. Additional airtime costs $0.10 per minute. The plan also includes "first incoming minutes free." This customer is paying $3.25 for "handset replacement insurance." If the customer damages the phone, Telephone Company D will replace it. The replacement plan may carry a deductible.

This customer has a couple of options to reduce this monthly bill. First, the customer is paying for the 500-minute plan but only used 291 minutes. If Telephone Company D has a 250-minute plan for $25, the bill could be reduced by about $20. Another way to reduce the bill is to cancel the $3.25

Telephone Company D

customer	account number	billing period ending	invoice date	page
Acme Manufacturing Joe Smith	00031965796-9	Oct. 4, 2001	Oct. 5, 2001	1 of 3

Account Summary

Previous Balance	123.70
Payments	-123.70
Current Activity Charges	53.25
Taxes	8.40
Total Amount Due by Apr. 1	**$ 62.65**

Retain For Your Records

Check Number	Date	Amount Paid
		$

Telephone Company D

To connect with Customer Care Dial 1-888-555-5555

Figure 20.3(a) Sample mobile telephone bill: page 1.

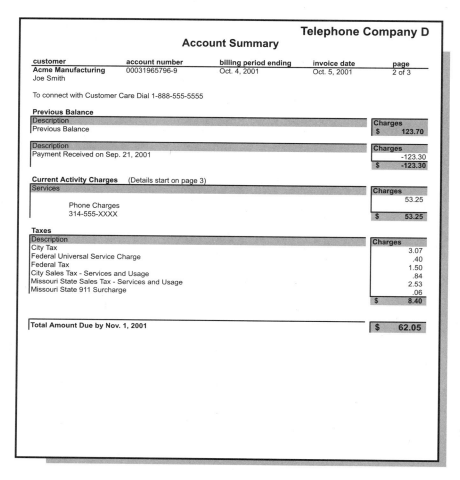

Figure 20.3(b) Sample mobile telephone bill: page 2, account summary.

monthly fee for "insurance." Mobile phones rarely need repairs; so as long as the customer is not tough on the phone, this plan is a waste of money. The customer also had 60 minutes of calling to an 800 number. This air-time could be eliminated altogether if the user would use a landline phone, such as a payphone. If the calls were made while driving, this option is not feasible.

Chapter 21 explains rate plans and airtime and offers numerous cost management strategies for these services.

Telephone Company D

Phone Charges

customer	account number	billing period ending	invoice date	page
Acme Manufacturing Joe Smith	00031965796-9	Oct. 4, 2001	Oct. 5, 2001	3 of 3

To connect with Customer Care Dial 1-888-555-5555

Phone Charges for 314-555-XXXX

Oct. 5 to Nov. 4
♦ $50.00 Monthly Service Charge
♦ Includes 500 Minutes To Use Anytime
♦ Includes Long Distance
♦ First Incoming Minute Free
♦ Caller ID, Call Waiting, 3-Way Calling
♦ Voicemail

Airtime Summary

Description	Minutes
Minutes Used in the Plan	291.0
First Incoming Minutes Free	0.0
Total Minutes	**291.0**

Call Detail

Date	Time	Phone Number	Call Destination	Rate/ Type	Minutes	Airtime Charges	Long Distance	Other Charges	Total Charges
Tue. Sep 18	8:45 AM	800-555-XXXX	800 Svc		37.0	included	0.00	0.00	0.00
Fri. Sep. 21	9:25 AM	314-555-XXXX	St. Louis, Mo		25.0	included	0.00	0.00	0.00
Sat. Sep. 22	8:35 AM	314-555-XXXX	St. Louis, Mo		3.0	included	0.00	0.00	0.00
Sun. Sep. 23	7:21 AM	314-555-XXXX	St. Louis, Mo		29.0	included	0.00	0.00	0.00
Mon. Sep. 24	6:52 AM	314-555-XXXX	St. Louis, Mo		61.0	included	0.00	0.00	0.00
Sun. Sep. 30	9:01 AM	314-555-XXXX	St. Louis, Mo		2.0	included	0.00	0.00	0.00
Sun. Sep. 30	9:06 AM	314-555-XXXX	St. Louis, Mo		21.0	included	0.00	0.00	0.00
Mon. Oct. 1	5:24 AM	314-555-XXXX	St. Louis, Mo		27.0	included	0.00	0.00	0.00
Tue. Oct. 2	6:21 AM	314-555-XXXX	St. Louis, Mo		34.0	included	0.00	0.00	0.00
Tue. Oct. 2	7:33 AM	800-555-XXXX	800 Svc		3.0	included	0.00	0.00	0.00
Thu. Oct. 4	5:01 AM	314-555-XXXX	St. Louis, Mo		12.0	included	0.00	0.00	0.00
Thu. Oct. 4	8:41 AM	800-555-XXXX	800 Svc		32.0	included	0.00	0.00	0.00
Thu. Oct. 4	9:51 AM	314-555-XXXX	St. Louis, Mo		5.0	included	0.00	0.00	0.00
Total					291.0	$0.00	$0.00	$0.00	**$0.00**

Phone Monthly Service Charges

Description	Charges
Promotion - Oct. 5 to Nov. 4	$ 50.00

Other Phone Charges

Item	Total
Handset Replacement Insurance	$ 3.25

Current Activity Charges for 314-555-XXXX	**$ 53.25**

Figure 20.3(c) Sample mobile telephone bill: page 3, phone charges.

21
Mobile phone rate plans

Wireless telephone service expenses include monthly recurring charges and the initial start-up costs. Start-up costs may include the cost of a new phone, the first month's access, and a one-time enrollment fee. Access is charged one month in advance, so this is always paid at the time of activation. Enrollment fees are designed to cover the carrier's initial administrative expenses. But in today's competitive marketplace, customers should insist that carriers waive this expense.

In the best-case scenario, the customer gets a free phone, pays no enrollment fee, and only pays for the first month's access. Prepaid cellular, on the other hand, normally has a high initial cost and the user must purchase a phone. Prepaid cellular does not include a monthly bill, however.

Free wireless phones

Carriers will normally give the phone for free if the user signs a term agreement. New customers should always ask the carrier for a free phone. The carrier will require at least a 12-month term agreement, which is negligible because most people will use the phone for at least 12 months. This is a win-win situation because the user gets a free phone, and the carrier can count on 12 months of revenue from the customer.

Besides term agreements, new users can cut their start-up costs by purchasing a phone apart from their carrier. Phones can be purchased over the Internet or at electronics shops, and, in many cases, a second-hand phone may be sufficient. In a corporate setting, employees regularly relocate or resign but company-owned cell phones are rarely recycled. A company can save $250 by recycling a phone instead of purchasing a new one for each new employee. Prior to purchasing the phone, be sure your carrier will program your phone and allow it to be used on its network.

Monthly billing for wireless telephone service

The typical cellular phone bill contains three charges:

- *Access:* A monthly recurring fee the user pays to use, or "access," the carrier's wireless network. Access is charged one month in advance.

- *Airtime:* Charges for the calls made over the past month. Airtime is billed after the calls are made.

- *Taxes and Fees:* Fees charged in advance for extra features, such as caller ID and voice mail.

A user who receives the bill on May 1 might be billed $35 for access in the month of May and $50 airtime for calls made in April. To use the carrier's network, you must pay for access in advance. Once you have made the calls, you are billed for the usage.

When users cancel wireless service, they are issued a small refund. Since they have paid for future access, but will not use it, the carrier refunds the money.

Rate plans

In order to balance their customers' needs with their own need to be profitable, wireless carriers have designed numerous rate plans for light users, heavy users, and corporate users. Neglecting to fine-tune their rate plans may make the difference between a slim profit margin and no profit at all. For these reasons, wireless phone rate plans change many times each year.

For consumer and corporate users alike, it is difficult to keep up with the plans and ensure each phone is on the most cost-effective rate plan. A business that diligently manages its wireless billing each month will regularly pay 20% less than a similar company that rarely examines its cellular billing.

Although wireless telephone rate plans regularly change, the plans' designs remain consistent. The following section explains the most common rate plans. Table 21.1 shows the four basic types: emergency plans, individual plans, small group plans, and corporate plans. These plans are billed monthly, but many of them are available as prepaid wireless plans.

Emergency plan

A typical emergency plan costs $15 to $20 per month for access, includes only 10 minutes of free calling, and has a high per-minute rate for additional minutes. Additional minutes may cost as much as $0.95 each. This

Table 21.1   Typical Rate Plans Offered in the Chicago Market

RATE PLAN	ACTIVATION FEE	ACCESS	INCLUDED MINIMUM	ADDITIONAL AIRTIME
Emergency				
AT&T Wireless $19.99	$25.00	$19.99	60	0.40 anytime
Individual rate				
Verizon Wireless "Digital Choice 250"	$35.00	$29.99	250 anytime	0.35 anytime
PrimeCo "Now Plus 400"	$25.00	$39.99	400 anytime	0.30 anytime
Sprint PCS "Free and Clear 1000"	$34.99	$99.99	1,000 anytime	0.35 anytime
Small group				
Cingular Wireless "Family Talk 1200"	$20.00 per line	$100.00	1,200 anytime, 500 weekend	0.35 anytime
Corporate				
Nextel "National Plan 4000"	$50.00 for 2 or more phones	$399.99	4,000 anytime	0.25 anytime

plan is designed for people who will only use the phone in the event of an emergency. It is ideal for workers who will only use the phone for one or two calls a month, such as security guards.

Many emergency-rate-plan customers put the phone in the trunk of their cars to be used only in case of a flat tire or other emergency. They can save money by canceling the monthly service, but keeping the phone. Any mobile phone will dial 911, even if the phone is not activated with a carrier.

If a caller uses the phone more than 10 to 30 minutes each month, he should change to an individual rate plan to avoid the high per-minute rates of the emergency plans. It is not unusual for an employee to grab a back-up cellular phone without changing the rate plan. In auditing the bills of large companies, I have seen employees grab a back-up phone and use it without changing the rate plan. All the minutes are then billed at very high rates. The company could have saved more than $100 per month by upgrading to a better rate plan.

Individual rate plans

Individual rate plans make up about 75% of all wireless telephone rate plans. These plans are tiered according to the number of airtime minutes included in the plan. Carriers normally require a term agreement with these rate plans, but during the term, they allow the user to upgrade or downgrade their plan to the next higher or lower plan.

Downgrading Rate Plans

For example, a traveling businesswoman chose a high-level rate plan and signed a 12-month term agreement with PCS PrimeCo. After 5 months, she decided to drop down to the next lower-level plan because she was not using all of the free airtime minutes. The carrier agreed to downgrade the calling plan, but required a new 12-month term agreement to start immediately. An additional 12 months were not added to her original 12-month term. Her total time commitment ended up being 17 months, not 24 months. Table 21.2 shows the impact of downgrading a rate plan. The calculations are based on the sample phone bill in Figure 20.3.

Upgrading rate plans

In the previous example, the customer chose a rate plan that was inappropriate for her call volume. She selected too large a plan and ended up

Table 21.2 Downgrading a Rate Plan Can Save Money

	Current Plan *"Dime Anytime"*		New Plan *"250 Plan"*	
	Minutes	Dollars	Minutes	Dollars
Access		$50.00		$25.00
Included airtime	500		250	
Actual airtime used	291		291	
Airtime cost	0.10	$ —	0.40	$16.40
Phone insurance		$3.25		$ —
Total:		$53.25		$41.40
Monthly savings:				$11.85
Annual savings:				$142.20

wasting money. The opposite scenario is just as common. Many wireless customers select a rate plan that is too small. They use all the free airtime minutes early in the billing cycle and then rack up airtime charges for the additional minutes. Upgrading to a higher plan will inevitably cut their cost.

One of the most common scenarios I have seen deals with office workers who go out as field representatives for their company. One such office worker never used more than his allotted 150 minutes each month. But once he became a field sales manager, his usage increased to 600 minutes each month. After his first month in the field, he upgraded to a higher rate plan. Table 21.3 illustrates the change.

Select a carrier that automatically changes your rate plans

The above examples represent the majority of wireless phone users who never seem to get their phones on the most cost-effective rate plans. One month their usage is up, the next month it is down. To avoid overpaying, they must vigilantly audit their bills each month. Today, some carriers automatically adjust a customer's rate plan to the most cost-effective rate plan each month. Their billing system selects the most cost-effective rate plan before printing the bill each month.

Table 21.3   Upgrading Wireless Rate Plans

	CURRENT PLAN "Single Rate 150"		NEW PLAN "Single Rate 500"	
	Minutes	Dollars	Minutes	Dollars
Access		$35.00		$49.99
Included airtime	150		500	
Actual airtime used	600		600	
Airtime cost	0.40	$180.00	0.35	$35.00
Phone insurance		$3.25		$ —
Total:		$218.25		$84.99
Monthly savings:				$133.26
Annual savings:				$1,599.12

Last-minute rate-plan changes

Some carriers allow "last-minute" rate-plan changes. Here's an example of how this works.

A small construction company has 10 mobile phones. The billing cycle ends on the 20th of each month. The office manager calls the carrier on the 18th or 19th and finds out how many minutes of airtime each phone had used. If a particular phone has used too much or too little airtime, she has the carrier change the rate plan. When the bill cuts the next day, the 20th, the new rate plan is in force, even though the calls were already made. This is possible because the billing system and the system that tracks airtime are independent of each other.

The prorated wireless telephone bill

When a rate plan is changed in the middle of the bill cycle, the next phone bill is a confusing mess. The bill reflects two minibilling cycles that are each half a month long. The first half of the bill shows a refund of unused access. The second part of the bill shows charges for half a month's access, and then another full month's access at the new rate.

Some carriers still do manual data entry, and they may neglect to give you the refund for the partial month. After a brief explanation, you should be able to negotiate this credit.

Why is it so hard to change rate plans?

Analysts and marketers design rate plans, which are then passed on to customer service and the billing department. By the time consumers become aware of the latest rate plans, the analysts are already drafting even newer plans. Consequently, customer service representatives often have outdated information about their own rate plans.

The primary job of a customer service representative is to keep busy talking to customers on the phone, rather than to study the employer's latest service offering. Customers can often get rate plans on the Internet days before customer service representatives are even aware of the new plans. Sometimes, representatives offer a rate plan that they cannot implement because the plan is either too new and not yet available or too old and no longer offered.

When they try to enter the change, their computer rejects it because the plan is not available. Instead, they arbitrarily choose a different rate plan for you. Consider yourself lucky if they call you back to tell you what happened. Otherwise, you will just have to see it on your next invoice—if you examine it.

Changing the rate plan on a cell phone seems like an easy task, but many things can go wrong. Human error or computer error can ruin the change. A simple typographical error will cause the wrong rate plan to be implemented and can cost you hundreds of dollars. What is intended to save a little money each month may actually raise your cost and waste a lot of your time.

How to ensure you have the best rate plan

I have used the following steps to reduce wireless phone bills hundreds of times for businesses throughout North America.

> *Get your historic average.* Look at your last 3 to 6 months' phone bills and determine your average amount of usage each month. Write down the amount of home airtime, roaming, and long distance. If your bills are unavailable, your carrier can provide this information.

> *Learn the newest rate plans.* Consult the Internet or your customer service representative to learn the new rate plans. Sometimes, your actual cell phone bill may have a notice that lists some of the new

plans. You may also see them advertised in your newspaper. It is also possible to compare rate plans side-by-side on the Internet (see Figure 21.1).

Do the math. The customer service representative will recommend the rate plan that is the most cost effective for you. His computers will do the calculations automatically. Always double-check the math, however, because the representative may have done the analysis on his own. Sometimes representatives accidentally recommend the wrong plan.

Change the rate plan. Ask the representative to change your account to the new plan. Make note of the representative's name and phone number. If the order fails to process, you are more likely to get a refund credit if you can prove that you did actually speak to a company employee. Have the representative fax or e-mail written confirmation to you that the order has been completed.

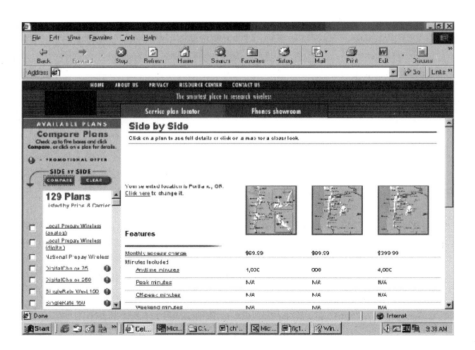

Figure 21.1 Side-by-side rate plan comparison on the Internet.

Confirm the change. After 2 to 3 days, call the carrier and ask for confirmation that your account has been changed to the new rate plan. Try to speak with a new representative who will objectively review your account. Avoid telling the company what plan should be in place; instead, have the representative first tell you what rate plan is in place. This is the most effective way to confirm the change.

Once you change plans, you can't go back

Watch out for grandfathered rate plans. If your cellular phone is on an old plan and you are considering changing to a newer plan, be careful. If your old plan is no longer offered by the carrier, you can never change back to it should you change your mind. In some cases, the older rate plans are better than what is currently offered.

Promotions

Wireless carriers regularly set aside funds to offer promotional deals to their customers. Promotions are designed to drum up new business for the carrier and are only for new customers, not existing customers. (Loyalty is rarely rewarded in the telecommunications industry.) If a new customer activates a phone with the carrier, she qualifies for the promotion. Promotions may include the following offers:

- Free night and weekend calling for a year;
- A free phone;
- A free battery;
- Free merchandise or gift certificates;
- Extra airtime minutes each month.

Promotions are usually advertised on the Internet, on radio, and in newspapers, but you can normally find the latest promotions by calling customer service. If you are an existing customer, you can still request the promotion. Corporate accounts rarely qualify for consumer promotions, but the corporate account executive may be able to pull some strings because the purpose is to retain the corporate account and develop more business from it. Many account executives are skilled in securing preferential treatment for their corporate customers.

Upgrading from analog to digital

In the early days of wireless phones, everyone used analog phones. In the late 1990s, customers began migrating to digital wireless service. Around 2000, the number of digital users equaled the number of analog users, with about 45 million of each type. The two main reasons for this trend are that digital service is higher quality and is more affordable. When digital service was first made available to the public, carriers offered very attractive pricing. The carriers had invested in building their digital networks and were eager to build their customer base.

The downside of switching from analog to digital is that you must buy a new digital phone, your coverage area may be different, and you may have to get a new phone number.

Save money by going back to analog service

For some wireless users, digital service is not the best option. Carriers offering analog service are still hungry for business and still offer competitive pricing. Of course, the rate plans vary from market to market, but especially in smaller, nonurban areas, analog service rate plans are usually the best. In these smaller markets, digital wireless service only has limited coverage, so a customer's analog phone might be more useful.

Small group rate plans

In a household or small business, multiple people may share the same cellular phone. Whoever is on the road grabs the phone on the way out the door. They would like to get another phone, but they do not want to pay the high access charge each month. Wireless providers are aware of this trend and have created small group plans to encourage these people to activate another phone or two.

A typical small group rate plan allows the customer to add a second or third phone and only pay $10 a month for access. The additional phones have no airtime allotment; their airtime is pooled with the first phone's airtime. By offering these plans, carriers sacrifice the earnings from access fees, but hope to increase their revenue with the extra call time.

This type of rate plan is often called a partner plan, family plan, or spouse plan. Carriers usually allow no more than four people to participate in a small group plan.

Converting to a small-group rate plan

Most businesses do not have enough cellular phones to qualify for a corporate rate plan, so each of their phones is on an individual rate plan. It is possible, however, to combine some of the phones under a small group rate
plan. It usually works best to partner a high-volume user with a low-
volume user. The key is to learn what plans are available in your market.
Carriers normally allow you to combine existing phones under a new
group plan. If one of the phones is currently under a term contract, you
might not be able to convert to a small group plan. Table 21.4 shows the
cost impact of putting two users together under a small-group rate plan. A
business with 10 or more phones may need to start multiple phone groups.

Table 21.4 Benefits of a Single Small-Group Rate Plan

	CURRENT PLAN: TWO INDIVIDUAL RATE PLANS		NEW PLAN: SMALL GROUP PLAN	
	Minutes	Dollars	Minutes	Dollars
AT&T Wireless $19.99				
Access		$19.99		$59.98
Included airtime	60		500	
Actual airtime used	70		420	
Airtime cost	0.40	$4.00	0.35	$ —
Total:				$23.99
AT&T Regional Advantage				
Access		$39.99		
Included airtime	200			
Actual airtime used	350			
Airtime cost	0.30	$45.00		
Total:		$84.99		
Total:		$108.98		$59.98
Monthly savings:				$49.00
Annual savings:				$588.00

Corporate rate plans

Most accounts only have one cell phone listed on the account. Customers typically activate only one phone at a time, and the phone is handled individually. Because of this, wireless carriers are not exactly sure what percentage of their customer base is commercial and what percentage is consumer. If, however, a business has multiple phones with the same carrier, it may qualify to be treated as a corporate account. Each carrier has a minimum number of phones to qualify for a corporate account. It may be as few as 5 or as many as 50.

If a business does not meet the minimum number of required phones, the carrier may make an exception and allow it to start a corporate account anyway. If a business is only one or two phones short of the minimum, it might cut its overall expenses by adding extra phones to qualify for the corporate account.

What is the advantage of a corporate account?

The main advantage of corporate accounts is that the monthly access charge is significantly reduced. Instead of paying hundreds of dollars a month in access charges, corporate accounts typically only charge $15 per phone. The corporate account for AT&T employees has a phenomenally low $10 monthly access fee and airtime only costs $0.10 per minute.

With individual rate plans, it is a gamble each month whether or not the user will use too many, or too few, minutes. In either case, money is wasted. Corporate accounts do not experience this waste. With a corporate account, high users pay a low rate for each minute, and low users only pay the minimal access fee. There are no surprises.

Pooled or not pooled?

Like small group accounts, some corporate accounts use a pool system for the airtime. A typical pooled corporate account may charge $100 for the first phone and $15 for each additional phone. A pool of 1,000 airtime minutes for all the phones to share is included in the plan. The customer must pay for any additional minutes above the first 1,000.

Nonpooled corporate accounts charge the customer a low access fee per phone and bill the customer for each minute of airtime. You probably will not be given a choice between pooled or not pooled, as most carriers only offer one or the other.

Using someone else's corporate account

Some carriers allow two businesses to combine their mobile phones to qualify for a corporate account. As long as the carrier separately bills the two companies, this is a great situation. I have seen large companies combine with their subsidiaries, customers, suppliers, and even employees to qualify for a larger corporate account. A large business can use its leverage to help a smaller sister company qualify for corporate pricing.

Avoid fraud and waste on corporate accounts

Some businesses allow their employees to put their personal phones on the corporate pricing. As long as the company does not pay for personal phones or a sister company's phones, this system works great. Larger businesses that do not routinely track their cellular phones and regularly audit their bills may end up paying for an employee's personal phone. It is not uncommon for an ex-employee to continue using the company-issued cell phone for months after employment has been terminated.

If you are auditing your bills for the first time, find out the name of the user for each cell phone listed on the bill. Some carriers print the user's name right on the bill. After reviewing your list of phone numbers and employee names, company department managers should be able to tell you which employees should be using company wireless phones. If you still cannot determine who the user is, call the number and see who answers. If all else fails, temporarily disconnecting the service should flush out the user. Be careful, because you may be surprised to learn whose phone you have canceled. I have canceled the phones of ex-employees, ghost employees, high-level executives, the college-aged children of executives, and even a high-level executive's mistress.

Before disconnecting anyone's phone, be sure your company's executive staff are well aware of the pending cancellation. On an account with 25 or more phones, this type of common-sense audit typically turns up one or two illegitimate cell phones.

Remove high-end users from the corporate account

Customers that use corporate accounts will have one or more phones that stand out because they use more airtime than the other phones. A typical corporate account has 20 phones that each use about 100 minutes of airtime each month and one or two phones that use more than 500 minutes of airtime. In this scenario, the larger users could be pulled out of the corporate account and handled as individual accounts.

Prepaid wireless

The four rate plans described (emergency, individual, group, and corporate) all require customers to pay for access one month in advance. The calls are charged at the end of each billing month, so each phone bill contains charges for the previous month's usage and the next month's access. Prepaid wireless users never pay a fee for access. To begin service, a new prepaid wireless customer simply buys the phone and pays for a block of airtime, such as 500 minutes. Some carriers charge an activation fee, but most do not; they make up for this lost revenue by requiring users to buy the phone directly from them. Some advantages of prepaid wireless are that the carrier requires no term contract and does not perform a credit check on a new customer. In Asia, prepaid wireless service is the only offering.

When additional minutes are needed, the user must buy a card and input the data into the phone. Most prepaid wireless systems have time limits. For example, at the end of six months, the phone is deactivated unless the user buys additional airtime. Once the additional airtime is purchased, any minutes that remained on the old card will again be available to the user. The Tracfone prepaid cellular program in the United States requires users to buy new cards every 2 months. If the customer fails to buy a new card, the phone is deactivated and the phone number may be reassigned.

Save money with prepaid wireless

Prepaid wireless is best for people who only sporadically use the phone. If a user regularly has months with almost no calling, he should consider prepaid wireless to avoid paying an access fee in a month when he will not be using the phone.

Association discounts

Like other telecom services, such as long distance, some wireless carriers offer association discounts. Being a member of a specific organization such as a chamber of commerce, civic organization, or trade organization may qualify your business for an association discount. Regional wireless carriers are more likely to offer association discounts than nationwide carriers. A cellular company operating near the automobile factories in Detroit gives a 5% discount off the total cellular bill to businesses that belong to a certain manufacturing trade organization. To find out about association discounts, ask customer service or one of the carrier's outside sales representatives.

Save money on features

The original cellular phones were cumbersome and came with no bells or whistles. Today, a host of advanced features are available. The most common features include the following:

* Call forwarding;

* Call waiting;

* Caller ID;

* Basic 911;

* Voice mail;

* Phone insurance.

Sometimes, these features are given to the customer at no charge, but other times carriers charge for the features. Telecom companies are notorious for nickel-and-dime charges that can double a monthly telephone bill. Sales representatives are given bonuses based on the number of advanced features they can sell.

The features are provided by the carrier's network technology. Once the features are available in the system, they cost the carrier almost nothing. Consequently, every dollar spent on features is sheer profit for carriers. Ask the carrier to waive the charges for all features. If the company says its billing system will not allow it, ask for a one-time credit equal to the cost of the features for a year. If the carrier will not waive the charges, consider canceling any unnecessary features.

One of the most unnecessary features is mobile phone insurance. This covers repair or replacement of your phone if it is broken, lost, or stolen. This feature costs $3 to $5 per month. With the exception of construction workers, most mobile phone users are not too hard on their phones. Today's phones are hardier than the original phones, and the new smaller sizes make them less susceptible to damage.

Cloning

Cloning is making fraudulent calls on someone else's wireless phone account. The thieves use radio scanners at major roads to intercept a cell phone's electronic serial number (ESN) and mobile identification number (MIN). Sometimes, industry insiders sell ESNs and MINs to the thieves. This information is then reprogrammed into another cell phone. The thieves then make calls that are billed on the other person's account. They usually use the clone phone for up to a month. Once the billing cycle ends, the carrier, or the customer, notices the dramatic increase in calling volume and disconnects the phone.

Customers are not liable for fraudulent charges; carriers must write these charges off. Cloning and other cellular fraud costs carriers half a billion dollars each year. If your phone has been cloned, notify your carrier at once. You should dispute the charges, and the carrier will remove the charges from your account. To remedy the problem, you will probably be issued a new telephone number.

22

Airtime

Like calls made on landline phones, wireless phone calls are subject to different rates. The combination of three variables determines the exact rate of a call made on a wireless phone. Each call receives a specific rate based on the following:

- *The caller's location.* Is the caller in the home area, or is he roaming?

- *Location of the number called.* Is it in the home area or long-distance?

- *Time of call.* Is it made during peak or off-peak hours?

A basic understanding of these factors is necessary to manage your wireless phone expenses efficiently. Figure 22.1 demonstrates these concepts in the Philadelphia market.

The home area

In the 1980s, the FCC established 306 metropolitan statistical areas (MSA) and 428 rural service areas (RSA). Each area is a specific geographic area based on a demographic market. Each of these 734 markets was to have two carriers: the A-block carrier and the B-block carrier. In the beginning, a customer's home area matched the MSA or RSA. Today, the industry has consolidated through numerous mergers, and the home area offered by many carriers may include more than one MSA.

When rate plans specify how many minutes of airtime are included in the package, they always refer to home airtime. Figure 22.1 shows a home airtime call and a long-distance call in the Philadelphia market. Figure 22.2 shows roaming and roaming with long-distance charges.

When the caller is in the home area and calls someone who is also in the home area, the airtime is considered home airtime. For example, a caller in Vineland who calls someone in Philadelphia uses home airtime. A call within the home area is like a local call on a landline phone; toll charges do not apply.

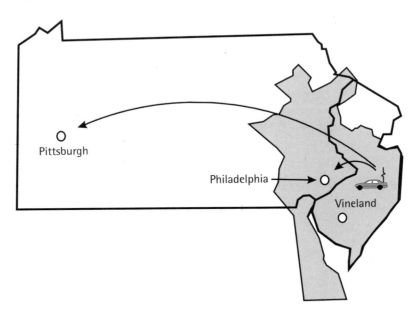

Figure 22.1 When the caller is in the home area, a call to Philadelphia is billed as home airtime, even if it is call between states. A call to Pittsburgh is billed as home airtime and long distance.

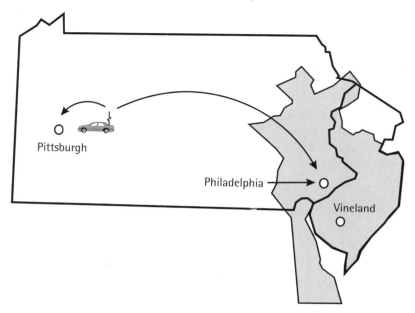

Figure 22.2 When the caller has roamed out of the home area, each call is billed the roaming rate. A call back to Philadelphia from Pittsburgh is billed roaming airtime and long distance.

Roaming

When the mobile phone user physically leaves the home calling area and uses his phone, his per-minute rate will be the roaming rate. He has roamed out of the home area and is now subject to higher rates. Roaming rates are typically $0.35 to $0.95 per minute. Sometimes a daily fee of up to $3 applies when a person roams. The rates are higher because the caller is borrowing the network of another home area.

Every wireless phone call is billed either at home airtime rates or roaming rates, but never both at the same time. The physical location of the caller determines whether his call is billed as home airtime or roaming airtime. If he is in his own home area, he is billed at the regular home airtime rates; if he is in a different market, he is billed the higher roaming rate.

If a caller has high roaming and long-distance charges, he should consider moving to a one-rate plan or using the dual numeric assignment mode (NAM) feature of his phone. These two ideas are explained in the next section.

Eliminate roaming charges with a one-rate plan

In the late 1990s, nationwide wireless carriers such as AT&T and Verizon began offering one-rate plans. With a one-rate plan, the user pays the same rate for all calls. It does not matter where the call originates or terminates. Both roaming charges and long-distance charges are eliminated. The entire country becomes the home calling area. Table 22.1 shows some sample one-rate plans.

One-rate plans are designed to be profitable because customers would roam in the same carrier's other markets. For example, a Miami-based AT&T customer would normally pay roaming charges in Chicago—but not with the one-rate plan. Because AT&T has a network in Miami and Chicago, the plan is profitable for AT&T. When the same customer roams in a market where AT&T has no network, AT&T must pay high roaming fees to the host carrier, but it cannot bill its customers the roaming rate. For this reason, one-rate plans were not as profitable as hoped. Consequently, many carriers no longer offer one-rate plans.

Wireless phone users who travel frequently and accrue significant roaming and long-distance charges should consider converting to a one-rate plan, which usually uses digital phones. When the user travels to an

Table 22.1   One-Rate Plans

Rate Plan	Activation Fee	Access	Included Minutes	Cost of Additional Minutes
Light user				
Cingular Wireless "Cingular Nation 100"	$15.00	$29.99	100	$0.35
Verizon Wireless "Single Rate 150"	$25.00	$35.00	150	$0.40
Medium user				
Nextel "National Plan 400"	$25.00	$59.00	400	$0.35
Verizon Wireless "Single Rate 400"	$35.00	$55.00	400	$0.35
Heavy user				
AT&T Wireless "Digital One Rate"	$25.00	$199.99	2,000	$0.25
Sprint PCS "Free and Clear 2000"	$34.99	$199.99	2,000	$0.35

area where digital service is not available, the phone switches to use the existing analog network in that area. When a digital phone is used in an analog area, additional charges may apply.

For example, a manager for an oil drilling company in Houston had a cellular phone through a local Houston cellular provider. His home area consisted of the metropolitan Houston area. On his weekly travels to the west Texas oilfields, he was out of his home calling area, and he accrued roaming charges. Calling back to the office in Houston, he was charged both roaming and long-distance charges. The manager changed to AT&T Wireless' one-rate plan and both roaming and long-distance charges were eliminated. Table 22.2 illustrates this change. In some remote parts of the West Texas oilfields, the manager could not receive AT&T's digital signal, so his phone defaulted to analog. In this instance, he was billed roaming charges for using the analog carrier's network.

Dual NAM

In another example, an account manager for a high-tech firm in Miami travels to Houston, where she is billed roaming charges. She decides to use the dual-NAM feature of her phone. Dual NAM means the phone is

Table 22.2   Benefits of a One-Rate Plan

	CURRENT PLAN "Cingular Home 450"		NEW PLAN "Cingular Nation 350"	
	Minutes	Dollars	Minutes	Dollars
Access		$39.99		$49.99
Included airtime	450		350	
Home airtime used	250		250	
Home airtime cost	0.45	$ —	0.35	$70.00
Roaming minutes	150		150	
Roaming cost	0.79	$118.50		Free
Long-dist. minutes	150		150	
Long-dist. cost	0.45	$67.50		Free
Total:		$225.99		$119.99
Monthly savings:				$106.00
Annual savings:				$1,272.00

capable of having two separate phone numbers. By using her phone's dual-NAM feature, the manager in Miami could continue to use the same cellular phone and phone number in Miami. In Houston, she would use a new phone number, but the same phone.

To take advantage of the dual-NAM option, the Houston cellular company must reprogram the phone. The phone then operates like an office phone with two phone lines. Line 1 is used in Miami and line 2 in Houston. When the manager moves between the two markets, she simply switches between her two numbers.

The main advantage of the dual-NAM feature is the elimination of high roaming charges. Table 22.3 shows this cost comparison. The downside is that the user will have two phone bills. Although it is expensive to maintain two accounts, it may be more cost effective than paying high roaming charges each month. Prior to using the dual-NAM function, research whether or not a one-rate plan would be more cost effective.

Long distance

Just like landline calls, long-distance calls on cellular phones cost more than local calls. Long-distance charges on a wireless phone generally cost $0.10 to $0.25 per minute. The boundaries are different, however. Cellular home areas might be 10 times larger than a landline local calling area. A person calling from Vineland, New Jersey, to Philadelphia is subject to long-distance charges on a landline phone. Because both cities are in the same home area, a call on a mobile phone is treated as home airtime and is essentially a local call (see Figure 22.1). But a cellular call from Vineland to Pittsburgh would be subject to long-distance charges in addition to home airtime charges.

Save money on wireless long distance

Cellular long distance is not as competitive as landline long distance, so callers do not have as many ways to cut the cost. Most users do have a few choices, however. First, call your carrier directly and ask for lower rates. Carriers usually have two or three different options. If you are still not satisfied, you can look into other long-distance carriers.

Many of the wireless carriers allow you to use a different company for landline long-distance. Your wireless provider can tell you which

Table 22.3   Benefits of the Dual-NAM Function

	CURRENT PLAN MIAMI HOME MARKET		DUAL NAM TWO HOME MARKETS	
	Miami Sprint PCS *"Free and Clear 400"*		Miami Sprint PCS *"Free and Clear 400"*	
Access		$49.99		$49.99
Included airtime	400		400	
Actual airtime used	325		325	
Airtime cost	0.40	$ —	0.40	$ —
Houston roaming minutes	150			
Houston roaming cost	0.69	$103.50		
Total:		$153.49		$49.99
			Houston Sprint PCS *"Free and Clear 180"*	
Access				$29.99
Included airtime			180	
Actual airtime used			150	
Airtime cost			0.40	$ —
Total:				$29.99
Total:		$153.49		$79.98
Monthly savings:				$73.51
Annual savings:				$882.12

long-distance carriers are available. For example, Ameritech Cellular allows its Michigan customers to choose long-distance carriers such as WorldCom to carry the calls. If you already use WorldCom as your landline long-distance provider, your cellular long-distance charges can be billed on your long-distance bill.

Caller ID and calling party pays

Few things are more frustrating than having a phone solicitor call you on your cellular phone. They read their sales script to you, and you have to pay for the call. Most digital wireless phones are able to use caller ID. Your phone displays the number of the caller when your phone rings. If you do not recognize the phone number, you can refuse to answer the call. Screening out unwanted calls is the main purpose of caller ID. By avoiding these calls, you do not have to pay for them. Another way to reduce the cost of inbound calls is to use a fairly new feature called *calling party pays*.

Calling party pays is a so-called advanced feature, and wireless companies charge a monthly fee of $2 to $5 a month for this service. Once you have signed up for calling party pays, callers cannot reach you by directly dialing your number. If they dial your normal wireless phone number, they will hear a recording that explains that you have enrolled in a caller-pays program. In order to reach you, they must hang up and redial using:

1 + area code + your mobile number

The additional dialing is intended to tell callers this is like a long-distance call—they will pay for it. The caller is normally billed $0.25 a minute, and the charges appear on the caller's local telephone bill.

Most businesses that use a lot of intracompany calling are better off without calling party pays. With the plan, their wireless bills are lower, but their local telephone bills will increase. With calling party pays, they pay $0.25 to call their own employees, who are in the field using wireless phones. If they cancel this plan, the call is billed on the wireless phone bill as home airtime. If the user has not exceeded the number of home airtime minutes included in the rate plan, the call is essentially free. Otherwise, the call will be billed under normal home airtime rates, which are always lower than calling-party-pays rates.

Full minutes or partial minutes

When shopping for a new wireless service provider, one should consider how the carrier bills the call time. Cellular calls have traditionally been billed in full-minute increments. A 2.5-minute call is billed as a 3-minute call. In the late 1990s, Nextel started billing in 1-second increments. The customer is only billed for two-and-a-half minutes for a 2.5-minute call. Most wireless phone calls are very brief, so the billing increment significantly impacts the actual monthly cost of a wireless phone.

Free first minutes

When PCS service was new, some carriers gave the first minute of a call at no charge. The free first minute was designed to stimulate more calling volume and, therefore, more revenue for the carriers. To find out the impact of the free first minute, simply look at the number of calls you made in a given month. The first minute of that call would have been free with a different carrier. Subtract the number of first minutes from your total airtime and recalculate the bill.

Free nights and weekends

Some carriers offer free nights and weekends for a flat fee of approximately $10 each month. Look at your call detail and add up the current cost of night and weekend calling. If it is regularly more than $10, you should sign up for this discount plan.

23

Paging service

Paging service was first offered in 1956. The pagers, made by Motorola, were large and only beeped, unlike today's tiny pagers that can receive both numeric and text messages. During the information revolution of the 1980s and 1990s, paging technology became more sophisticated, and pagers became smaller, more fashionable, and more affordable.

About half of the pagers in service today are used by consumers, not businesses. Owing to today's affordable prices, pager use is growing, especially among young adults and teenagers. Many high schools and junior high schools have banned pagers on campus. Some predicted that cell phones would make pagers obsolete, and many people have replaced their pagers, but the paging industry currently boasts an astonishing growth rate of 30% per year. More than 27 million pagers are in use in America today, and the industry earns almost $5 billion in annual revenue.

Paging service typically costs $3 to $50 a month, depending on four factors: contract discounts, type of service, coverage area, and billing cycle. The base rate of $10 a month for a rented digital pager is a fair representation of pricing in today's marketplace, so this figure will be used in examples throughout this chapter.

Level of service

The level of service provided by the paging carrier depends on the type of pager. The three types of pagers offered today are digital pagers, alphanumeric pagers, and two-way pagers. Some old-tone pagers are also still in use. Table 23.1 shows typical monthly pager rates:

Tone pagers

Tone pagers, or beepers, were initially used by people whose professions required them to be available at all hours, such as doctors. Tone pagers give off a tone only; they cannot send numeric messages. Tone pagers are definitely a telecom antiquity; it has been my experience that out of a hundred businesses, five or six still have tone pagers.

Save money on tone pagers

Ironically, tone pagers often cost more than digital pagers. During the past few years, competition has driven the cost of digital pagers down, but tone pagers are a small noncompetitive sector of the overall paging industry. Carriers and customers alike tend to ignore their tone pagers, and no one ever tries to negotiate lower pricing.

Replace tone pagers with digital pagers

Consider the following example: A West Texas oil drilling company used 20 tone pagers. When an employee out in the oilfield was paged, he knew it was time to return to the office. The system worked fine, and the company never felt the need to upgrade to numeric pagers. Since the 1970s, the company had been paying $15 per month for each pager. Eventually the company replaced the tone pagers with digital pagers for only $7 per unit. This change saved the company $160 per month.

Table 23.1   Typical Monthly Pager Rates

Tone pager	$10
Digital pager	$10
Alphanumeric pager	$15
Two-way pager	$35

Digital pagers

Digital pagers, also called numeric pagers, relay numeric digits to the user—normally, a phone number. Most pagers in service today are digital pagers. Digital pagers normally sell for $25 to $100, but carriers will often give the pager to a customer who signs a contract. Monthly service for a digital pager costs between $5 and $10.

Alphanumeric pagers

Alphanumeric pagers display numeric and text messages across a small LCD screen with up to nine lines. Many carriers broadcast news, weather, and stock quotes to the pager throughout the day. Alphanumeric pagers also come loaded with such features as distinct rings and vibrations, an alarm, message memory, calendar functions, and phone-number storage. They cost between $100 and $300 each. The expense for service is normally $10 to $20 per month.

Two-way pagers

Two-way pagers, such as Motorola's Page Writer, provide traditional alphanumeric paging capability, but also give the user the ability to check e-mail, surf the internet, and read voice-mail messages that have been converted to text messages. Two-way pagers look like miniature laptop computers and even include a small keyboard. These pagers cost around $400, and service costs between $25 and $35 per month.

Paging coverage areas

Paging service is offered in specific geographic areas. Unlike other telecom services, these areas are not dictated by government regulations. Paging service areas are only limited by the number of towers the paging company is willing to build. Paging providers usually divide their coverage offerings into three categories: local, regional, and nationwide. Table 23.2 shows typical monthly pager rates for a digital pager.

In metropolitan areas, customers can choose between numerous carriers, but in remote areas customers may have only one choice. In this case, there is little leverage to negotiate pricing. The paging company is the only game in town, and it can set its rates without being influenced by external factors, such as competition.

Table 23.2   Typical Monthly Rates for Digital Pagers

Local coverage	$10
Statewide coverage	$15
Regional coverage	$20
Nationwide coverage	$25

Local coverage

The most basic, and therefore most economical, paging area is called the local coverage area. Local coverage is usually the size of a small state or metropolitan area. Figure 23.1 shows Central Link's local coverage area in the Waco, Texas, market. The paging towers broadcast in all directions, which results in a circular pattern around the edges of the coverage area.

Statewide coverage

In large states, paging companies divide the coverage into two or three local areas. California is split into a northern and southern area. For the pager to work in both areas, the customer must pay for statewide coverage that costs

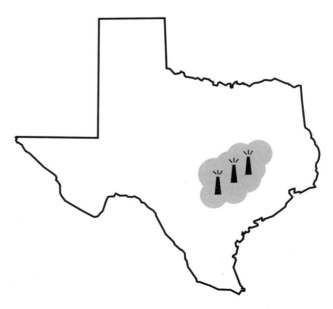

Figure 23.1 Local coverage area for paging in Waco, Texas.

a few dollars more than local coverage. Depending on the carrier, Texas has about seven local coverage areas. Paying for statewide coverage ensures that the pager will work in all seven areas. Figure 23.2 shows Page One's statewide coverage in Texas. Note that the pager will not work in some rural areas that are not reached by Page One's towers.

Regional paging

Most nationwide paging companies divide the country into four regions. Each region contains a dozen separate local coverage areas. This scenario is called regional coverage, and, of course, costs a few dollars more than local coverage. Figure 23.3 shows Arch Paging's different regions.

Nationwide paging

No paging supplier offers true nationwide paging. Nationwide means the pager will work in the nation's largest cities and usually along major interstates, but the service is choppy in remote areas. In 1983, SkyTel first offered nationwide paging, but today, many other carriers offer this service.

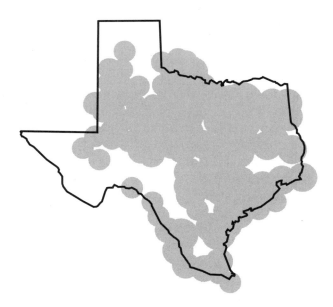

Figure 23.2 Statewide paging coverage map in Texas.

Figure 23.3 Regional coverage area for Arch Paging.

Paging bills

Paging bills are the simplest telecom bills to understand. The bill shows each pager number with an itemized list of additional charges. These four charges are usually listed for each pager: equipment rental, service, maintenance, and usage.

Some paging carriers do not itemize their charges. They just give one line item for each pager, in which case the charge is normally for equipment rental and service. Some Mom-and-Pop carriers simply show the total charge on one line, such as "pager service for 26 units ... $260." The customer should always request an itemized bill in order to verify all of the charges.

The sample bill in Figure 23.4 shows that the customer is paying $5 per month for paging service and another $5 per month for equipment rental. This customer could save $10 per month by purchasing the pagers instead of renting them. One of the pagers has an 800 number that is being billed indirectly by the paging company. The bill also notes "001 Pager Contract," indicating a volume or term contract has been signed. The bill gives no

DATE	DESCRIPTION	SERVICE		EQUIPMENT		OTHER		TOTAL	
	Acme Manufacturing **100 Main Street** **Memphis, Tn XXXXX**					**Pager Company A**			
9/15/01	BEGINNING BALANCE							31	82
9/30/01	PAYMENT **THANK YOU**							31	82
	BALANCE FORWARD								00
	CURRENT CHARGES								
	001 Pager Contract								
10/15/01	LOCAL DIGITAL 901-821-XXXX 10/15/01-11/15/01 Pager Protection	5	00	5	00	2	00	12	00
10/15/01	LOCAL DIGITAL 901-821-XXXX 10/15/01-11/15/01 Pager Protection PERSN 800 Access	5	00	5	00	2 5	00 00	17	00
10/15/01	FINANCE CHARGES								00
	STATE TAX							2	32
	UNIVERSAL SERVICE CHARGE								50
10/15/01	TOTAL CURRENT CHARGES							31	82
10/15/01	TOTAL AMOUNT DUE							31	82

** THE 800 ACCESS PORTION OF YOUR BILL IS PROVIDED AS A SERVICE TO THE ABOVE CARRIER.
THERE IS NO REQUIREMENT YOU CHOOSE THIS CARRIER FOR ACCESS SERVICE

PAGER COMPANY A
123 MAIN STREET
URBANA, IL XXXXX
800-555-5555

CLOSING DATE	INVOICE NO.	ACCOUNT NO.
10/14/01	351654651	316546132

Figure 23.4 Typical pager bill.

detail, but the carrier's customer service representatives should be able to explain the terms of the contract quickly.

Equipment rental

Some customers purchase pagers (usually from their carrier), but most customers rent them from their carrier. Businesses watching their cash flow may want to avoid the up-front cost of purchasing new pagers, but paging companies steer most of their businesses toward renting. Equipment rental for a local digital pager costs between $3 and $10 per month.

Rent or buy

One easy way to reduce monthly paging costs is to purchase pagers instead of renting them. The only real advantage of renting pagers is that carriers will repair damaged pagers at little or no cost to the user. If repairs are needed to customer-owned pagers, the customer has to foot the bill. But pagers are fairly low-tech and do not require much maintenance.

For example, a West Coast architectural firm rented 15 digital pagers from its carrier. Each month, the company paid $5 for rental and $6 for service. After performing a miniaudit of its telecom expenses, the company was surprised to learn it had spent more than $1,800 in rental fees over a 2-year period. When it first began using pagers, the company could have purchased all 15 pagers for no more than $750. It presented its findings to its carrier, which ended up giving the firm the pagers for free. Other paging companies may not have been as generous, but they would at least be willing to sell the pagers at a greatly discounted price.

Service

Paging providers bill for service in advance. This is the basic charge for paging; it covers the cost for the tower to transmit messages to the pager. The monthly service for a digital pager with local coverage is usually only $3 to $5.

PRP—Insurance for your pager

Pager maintenance programs, also known as pager replacement programs (PRP), and pager protection, function like an insurance policy. If the pager is lost, damaged, or stolen, the paging company will replace the unit for free. PRP for a digital pager usually costs $1 to $3 per month.

Save money on pager replacement programs

Consider this example: A northern California hospital used more than 100 pagers for its nursing staff. The hospital's large volume allowed it to negotiate a low rate of $5 per month for each digital pager, which included service and rental. The pager salesman persuaded the hospital telecom manager to purchase PRP for each unit at a cost of $1 per unit and a $25 deductible to replace a pager.

The manager later realized that the nurses were very gentle with the pagers. Only one pager was lost or damaged each month. To replace the one lost pager, the hospital had to pay the $25 deductible. After adding this $25 to the $100 already spent for PRP, the hospital was spending $125 each month to replace one pager it could have bought for only $50. The manager quickly canceled PRP and ended up saving the hospital more than $1,000 in the first year. Five nurses would have to lose their pagers each month before PRP was cost effective (see Table 23.3).

To determine whether or not your organization should keep or cancel pager protection, calculate your own break-even point. Then, find out how many pagers are lost or damaged each month; if you do not know, the paging company should be able to tell you.

Table 23.3 Pager Replacement Programs: Calculating the Break-Even Point

PAGERS LOST EACH MONTH	PRP COST ON 100 PAGERS	$25 DEDUCTIBLE FOR LOST PAGERS	MONTHLY COST WITH PRP	MONTHLY COST WITHOUT PRP
0	$100	$0	$100	$0
1	$100	$25	$125	$50
2	$100	$50	$150	$100
3	$100	$75	$175	$150
4	$100	$100	$200	$200
5	$100	$125	$225	$250
10	$100	$250	$350	$500

Companies that are rough on their pagers should keep PRP. A company of rugged construction workers who destroy half a dozen pagers each month may actually save money by using PRP.

Usage

Paging companies originally charged customers a flat rate each month, but over the past few years, this has changed. Today, some providers also charge as much as $0.25 per message, but they generally give the user the first 100 messages at no charge.

Paging billing cycles

Paging customers can choose to pay their bills monthly, quarterly, semiannually, or annually. Carriers offer price breaks to customers on quarterly, semiannual, or annual billing. The longer the billing cycle, the lower the price. With annualized billing, carriers have lower administrative costs, which include the costs of processing, printing, and mailing an invoice. They are willing to pass these savings on to customers because their own internal processes are streamlined. Table 23.4 shows typical pager pricing for a local digital pager.

Using the pricing shown in Table 23.4, a business with 10 pagers pays

10 pagers * $10 per month * 12 months = $1,200 per year

If the company converts to annual billing, it will only pay $800 per year. The hassle of receiving a bill and cutting a check each month is also

Table 23.4   Typical Pager Pricing for Digital Pagers

Monthly billing	$10
Quarterly billing	$25
Semiannual billing	$45
Annual billing	$80

eliminated. A customer can almost always cut its paging costs by shifting to annualized billing. But, if a pager is taken out of service during the year, make sure the pager company issues a prorated refund for the months that the pager will not be in use.

Save with term agreements

Like other telecom services, pager providers offer 12-, 24- and 36-month term agreements. The carrier benefits by locking in the customer's revenue for the entire term, and the customer benefits by receiving a lower monthly rate for each pager. Signing a 12-month term agreement with a paging supplier usually knocks $1 off the monthly cost per unit. A 24-month term agreement gives a $2 discount, and a 36-month agreement gives a $3 discount.

If the negotiated pager cost is already low, such as $5 per unit, a term agreement will probably not create any additional discounts. Paging is a low-margin business, and carriers make sure each transaction remains profitable for them.

A loophole with pager term agreements is that carriers sometimes fail to specify a minimum number of pagers. A business with 100 pagers that becomes dissatisfied with its paging company could theoretically move 99 pagers to another paging company and face no penalty with the original carrier.

Consolidate to one carrier

One of the most basic rules of telecommunications management is consolidation of services to one vendor. The majority of companies I have worked with use multiple paging vendors because they have no centralized control over their telecom services. Newly merged companies still use different vendors, and employees are often allowed to choose their own vendor. One single company site might have up to three or four separate paging providers. It is difficult for accounts payable and the telecom department to keep up.

For example, a leading managed-care corporation has more than 50 nursing homes. The company was aggressively buying and building new nursing homes at the rate of one per month. The new corporate telecom manager noticed the following trends:

- The staff at each nursing home used an average of five pagers.

- The corporation processed more than 100 separate invoices each month.

- Some invoices were for a single pager; others were for as many as 24 pagers.

- Ten different paging vendors were used.

- It took two full days each month just to process the pager bills.

- Although about a third of the pagers were with PageNet, the pagers were billed on a number of different invoices.

- The average cost for digital pagers was $11 per unit.

- Many of the accounts had pager protection, which cost between $1 and $2 per month.

The telecom manager decided to only keep the PageNet pagers and replace the others with PageNet pagers. PageNet assigned an account executive to the account, who immediately consolidated all billing into one bill that would be sent directly to the telecom department at the corporate office. Pager protection was canceled, and the large volume allowed the manager to negotiate a low price of $4 per unit.

In most cases, a national account is not cost effective, because pager companies cannot offer one price and one bill. Each region is run as a separate company and the pager companies can only offer price breaks based on the number of pagers in their own region.

24

Contract negotiation

Businesses that do not diligently manage their telecom expenses always pay too much. Telephone companies make a lot of money from customers who do not proactively manage their phone expenses. Most of this book offers telecom cost-management strategies from a *reactive* standpoint. The customer audits his bills, fine-tunes his telephone accounts, and takes action to reduce his costs. This chapter offers proactive cost management strategies for dealing with telecom contracts. Negotiating a new contract is the single most significant way for a business to cut its telecom costs.

This chapter explains the basic elements of a telecom contract and the most common special clauses that may be in a contract, then offers advice on how to negotiate a favorable contract with a telecom carrier. The information applies to all types of telecom contracts, including local service, long-distance, data, and wireless service.

The three phases of procuring telecom services are represented by the following documents:

+ The proposal;

+ The contract;

+ The phone bill.

The carrier first gives a proposal for services. A contract is signed. Then, a month later, the customer receives his first phone bill. To avoid being over-charged, the customer must give careful attention to each of these three phases. Only then can a business stay in control of its expenses. Phone companies are normally not out to deceive their customers, but their complex bureaucratic processes frequently put the customer in an unfavorable position. Telecom contract negotiation has many pitfalls that open up a business to undue financial risk.

Every customer's situation is unique. Service offerings and contracts vary from carrier to carrier. But some things remain consistent, and this chapter will explain the contracts and tactics most frequently used in today's marketplace.

The carrier's objective

Like other business contracts, the supplier's core objective with a contract is to protect its own interests. The telephone company's mission is to keep the client locked in and keep other carriers locked out. It locks the client in with numerous contract restrictions, such as volume commitments, term commitments, and exclusivity clauses. It locks other companies out with confidentiality agreements and by making the contract difficult to compare to those of other suppliers.

Long-distance carriers are especially notorious for having lengthy contracts that are almost impossible to compare to another carrier's contract. The average business has neither the expertise nor the time to study these contracts, so it usually just resigns with its current supplier or crosses its fingers, signs with a new carrier, and hopes that all goes well.

Typical contract outline

Most telecom contracts include discussion of the following:

+ Effective dates;
+ Tariff reference;
+ Services provided;

* Volume and term agreements;

* Discounts, promotions, and credits;

* Termination and penalties;

* Special conditions, special clauses, and addendums.

These are the most important elements of the contract from a cost-management perspective, but numerous other clauses may be included covering issues like payment terms, confidentiality, and jurisdiction.

Effective dates

This section of the contract explains when the new term starts. One contract states as follows:

> ... beginning with *the next billing cycle* following MCI WorldCom's execution of this agreement, customer shall pay the rates and receive the discounts associated with ... this agreement (MCI WorldCom On-Net Voice Agreement).

Customers expect to see the new rates on the next phone bill after signing a new contract with their telephone company. The above clause contains this promise, assuming MCI WorldCom does not delay in executing the agreement. Customers are always surprised when they receive two or three months of billing at the old higher rates before the new rates are in place.

When that happens, the sales representative that has to face the disgruntled client usually says, "We want to make sure your contract is implemented correctly, so it takes a while ...." In reality, the contract has probably been sitting on someone's desk for weeks.

Another contract states as follows:

> The term of this agreement is the period of time indicated below, commencing upon Sprint's acceptance and implementation of this agreement (Sprint Real Solutions Sales Agreement).

Once the customer signs the contract, Sprint has an unspecified amount of time to implement the new pricing. In other words, the carrier will implement this contract when it feels like it.

Some customers do not mind waiting for the carrier to implement the new pricing. But the main reason for signing a new contract is to get lower rates and save money each month. A typical new contract may reduce a customer's bill by 30% a month. Every month that the contract is delayed, the customer pays too much. When this happens, the customer should diligently fight for a refund of the overcharges. If the carrier hesitates to issue a credit, the customer may consider short-paying the bill by the disputed amount until the carrier resolves the problem.

There is another reason that delayed implementation is unfavorable to customers. A 36-month contract that is delayed 3 months essentially becomes a 39-month commitment for the customer. Carriers love this because they can count on 3 extra months of revenue on the back end of the term. This customer should require the carrier to change the effective date of the contract to when the contract was first signed and not the date that the contract was finally implemented.

Tariff reference

As mentioned in earlier chapters, a tariff is a document that discloses pricing for a specific telecom service. Phone companies are required by law to file tariffs with the government. Most customers who purchase a telecom service never see the actual tariff for that service.

The beginning lines of a telecom contract will almost always refer to a tariff, as does the following:

> The service and pricing plan you have selected will be governed by the rates and terms and conditions in the appropriate AT&T tariffs as may be modified from time to time (AT&T Business Service Simply Better Pricing Option Term Plan Agreement).

Phone companies spell out their pricing in the tariffs, not in the actual contracts. Contracts always state discount amounts, instead of specific rates. Every year, multibillion dollar corporations sign long-term telecom contracts that give no guarantee of the pricing. The pricing is negotiated during the initial proposal phase and, when the contracts are to be signed, the customer can only hope that the contract matches the promised rates.

Does the contract give a specific rate or just a specific discount?

In the rare case of a contract actually stating specific rates, the rate should be effective for the life of the contract. Normally, however, only the discount is specified. The customer will receive that discount for the life of the contract, but if the carrier raises the tariff rate, the customer's net effective rate will be higher. Chapter 12 gives a detailed explanation of long-distance rates and how they are affected by rate increases.

Services provided

This part of the contract lists the services the customer is purchasing from the carrier. Some contracts specify the exact service the customer is purchasing, as does this frame relay contract excerpt:

> This MCI WorldCom frame relay service agreement is made between ... MCI WorldCom and XYZ Company (MCI WorldCom Frame Relay Service Agreement Contract).

Most contracts, however, only list their specific product names, such as the following line from a Pacific Bell contract that gives a discount on local calling: "Customer elects to subscribe to Value Promise Plus Term Discount." A standard contract template used by Sprint for customers spending up to $960,000 per year simply states, "Customer subscribes to the real solutions (annual) integrated services selected below." The customer then selects the services by checking the appropriate boxes.

Volume and term agreements

Telecom term commitments are usually for 12, 24, or 36 months. In the case of Centrex and some data contracts, the term commitment may extend as far as 7 years. Carriers that are new in the marketplace, such as CLECs, rarely require long-term agreements because most customers are unwilling to be tied to an unknown CLEC for an extended period of time.

Volume agreements normally apply to measured telecom services such as local or long-distance calling. Volume commitments are either measured in prediscounted gross dollars or post-discounted *net* dollars. A California trucking firm, for example, had a $10,000 gross volume commitment with AT&T. Its gross billing was about $12,000 per month, but with its 45% discount, the net amount of the bill was only $6,600 per month. If the company's contract had a $10,000 net volume contract, then AT&T would require it to pay a shortfall penalty of $3,400 each month. It is extremely important for a business to understand whether its volume commitment is gross or net.

What services contribute to the volume commitment?

Today's so-called supercarriers can provide local, long-distance, data, and wireless service to a single business. If a customer moves all its services to one carrier, then all of the services should contribute to the volume commitment. Greater volume may allow the customer to qualify for a higher discount level.

All telecom services with the same carrier usually contribute to the volume commitment, but some carriers may exclude some services. The following excerpt from a Sprint contract lists the charges that will not be counted toward fulfillment of the volume commitment:

> Features, equipment, directory assistance, operator services and any other access charges … not specified above are *not contributory* to meeting the minimum annual commitment (Sprint Real Solutions (annual) Sales Agreement, Percentage off).

A Sprint customer with significant volume in any of these areas should attempt to have this volume contribute toward fulfilling the minimum annual commitment.

When is the true-up?

Volume commitments are either expressed monthly or annually. A $10,000 monthly commitment is not the same as a $120,000 annual commitment, especially for a seasonal business. Annual commitments pose less risk for a customer because extra volume in a high month may help to offset low volume in a different month. Seasonal businesses, such as summer camps and ski resorts, are better off with an annual-volume commitment.

As a general rule, businesses should only agree to a volume commitment that is 75% of their expected volume. This excess volume cushion will help the business avoid a shortfall penalty. It also creates more negotiating leverage with the carrier. Carriers are always afraid that customers will move their uncommitted overflow traffic to another carrier.

Discounts, promotions, and credits

This section of the contract lists a specific discount off the tariff rates. A contract may have multiple discounts, especially single contracts that cover multiple services, such as long distance and frame relay. The following contract clause lists a 10% discount for all services and then adds an additional 29% discount toward interstate calling:

Customer will receive an additional ten percent (10%) monthly discount on all domestic and international long distance usage charges and frame relay and private line charges associated with the on-net voice customer ID assigned hereto and inserted below by MCI WorldCom ... Customer will receive an additional twenty-nine percent (29%) monthly discount on all interstate long distance usage charges (MCI WorldCom On-Net Voice Agreement).

In addition to monthly recurring discounts, this section of the contract should also list promotions and credits. Promotions are added on to a basic contract. Customers are first offered a contract with a standard discount based only on the volume and term-commitment levels. Sometimes the standard discount is not enough to persuade a customer to sign a new contract. Telephone companies' branch managers can offer additional incentives to these customers. Every quarter, some carriers make funds available to be used to win more customers for the branch. The funds are given out in the form of promotions. Some examples of promotions include the following:

- An additional 10% discount on international calls;
- An additional 35% discount on local loops;
- A $1,000 invoice credit given in the 13th month of the contract.

Invoice credits

The most common promotions are invoice credits given to customers, such as a sign-on bonus that would appear on the first month's phone bill. Customers quickly forget the generosity of their carriers, so carriers have started issuing credits in the later months of the contract term. Today, a customer who signs a 36-month contract with MCI WorldCom might qualify to receive credits in the 13th, 24th, and 30th months of the contract. Stretching out credits into the latter end of the term also helps the carrier keep customers from renegotiating their contract too early.

Carriers also give credits to offset the expenses incurred by the customer to install new services. A customer who spends $5,000 on Cisco routers for a new frame-relay network might be given a $5,000 invoice credit by its frame-relay service provider to offset the cost of the routers.

Specific promotions must be negotiated prior to signing contracts. Even then, the final contract must be reviewed to ensure that the promised promotions are actually noted in the contract or in an addendum. If a promotion is left out of the contract—by accident or malice—the customer may never receive the promotion, even though it was discussed during negotiations.

Termination and penalties

A customer can terminate a telecom contract in one of two ways: without liability or with liability. Most contracts are terminated without liability.

Termination without liability

There are three ways to terminate a contract without liability.

Fulfilling the volume commitment early   At times, customers can switch carriers before their term commitment is up if they have fulfilled their volume commitment. Some customers in this situation have left a token phone line with the old carrier so they can still fulfill the term commitment. A customer who signs a 12-month, $120,000 contract with AT&T may be able to switch to Sprint in the eighth month if it has already paid more than $120,000 to AT&T.

The carrier's failure   Sometimes carriers fail to provide the contracted services. For example, a business whose dedicated private lines experience repeated down time may be able to cancel a carrier contract, especially if the contract contains a quality assurance clause. Customers who experience repeated billing errors may also be able to get out of a contract. Contracts never specify that the carrier is required to provide an accurate phone bill, but if a customer experiences many months of fouled-up phone bills, the carrier might let the customer out of the contract.

Out with the old   The simplest way to get out of a telecom contract is to replace it with a new contract. Of course, this can only be done with the same carrier and is normally only done toward the end of the existing term. But telecom contracts can be renegotiated at any time as long as the customer has leverage. A customer who is ordering additional services and is therefore increasing his volume is a prime candidate for renegotiating his contract. The following contract excerpt illustrates how an old contract can be replaced by a newer one:

> Customers may discontinue their Simply Better Pricing Option Term
> Plan prior to the expiration of its term without liability if they concur-

rently replace service under the plan with a newly subscribed AT&T plan that has a specified revenue commitment equal to or greater than the remaining revenue commitment under the plan being discontinued (AT&T Business Service Simply Better Pricing Option Term Plan Agreement).

AT&T is essentially telling the customer, "We'll give you a new contract, as long as we get more money or time out of you." If the customer gets better rates, it is a win-win situation. Before renegotiating, the carrier will consider the contract value of the customer. Contract value is a way carriers look at a customer's current contract and determine the financial value of that customer. See Chapter 13 for more detail on contract value.

Termination with liability

The AT&T Simply Better Pricing Option Term Plan contract explains how a termination with liability is handled:

> Customers who terminate their Simply Better Pricing Option Term Plan prior to the expiration of the selected term period will be billed a termination charge equal to the monthly revenue commitment multiplied by the number of months remaining in the term period... (AT&T Business Service Simply Better Pricing Option Term Plan Agreement).

Here, AT&T tells customers they can cancel the contract, but it's going to cost them. The customer must pay AT&T the remaining amount of the volume commitment, which happens to be the same amount of money AT&T would have earned from this customer anyway. If, however, the volume commitment is expressed in gross dollars, the customer pays more in a shortfall situation, because the shortfall amount is usually not discounted.

Special clauses

Most special clauses have their roots in previous lawsuits that disputed the exact meaning of a telecom contract. After the lawsuits, carriers beef up the language in their contracts to avoid similar lawsuits in the future. Consequently, telecom contracts have gotten rather lengthy. AT&T Uniplan contracts are rarely less than 10 pages long. (All those pages, and they still cannot specify an exact cost per minute!)

Like the carriers, customers also want to protect their interests, so they have written clauses of their own. As a result, some clauses are carrier friendly, while others are customer friendly. The most common special clauses are explained in the next section.

Special clauses that favor carriers

The following clauses are some of the most commonly encountered in a telecom contract that protects the telephone company's interests.

Exclusivity   Telephone companies are always afraid of losing customers. They know that if they lose a portion of a customer's traffic, they may end up losing all of the customer's traffic. The exclusivity clause requires a customer to use one carrier exclusively. The following excerpt shows Sprint's intent to control all of a customer's long-distance traffic:

> Customer will award Sprint 100% of its and its affiliates' long-distance telecom service. For minimum annual commitment levels equal to or greater than $300,000, customer will award Sprint 90% of its and its affiliates' long distance telecom service (Sprint Real Solutions (annual) Sales Agreement).

This clause clearly spells out that the customer must use Sprint exclusively as the carrier for all its long-distance service, unless its annual volume commitment is greater than $300,000. In that case, Sprint allows just 10% of the customer's traffic to be with another carrier. This clause is not easily enforced by carriers. After all, how can they tell what traffic another phone company is carrying?

Traffic requirements   Telephone companies have different profit margins for different services. A carrier may earn a lot of money from interstate calling, but have slim earnings from intrastate calling. Consequently, many carriers may include a clause in the contract that requires the customer's traffic to have a minimum percentage of interstate calling. This clause normally reads as follows:

> Customer agrees that 75% of its long-distance call volume will be interstate traffic…

The contract may give stiff penalties for not fulfilling the specified traffic requirements, so customers should always negotiate a lower minimum percentage or have this clause removed altogether. Carriers have no efficient process to enforce this clause, so·most customers who fail to meet the specific traffic requirements will probably not face any repercussions from the carrier.

Escalating MAC   The minimum annual commitment (MAC) is the amount of money the customer commits to spending with the carrier in a given year of the contract. The MAC may be expressed in gross dollars or net dollars. Most contracts have a level MAC for each year of the contract. Extremely large contracts are the exception to the rule, however. Very large telecom contracts with volume commitments in the neighborhood of $1 million per year may have an escalating MAC clause. This is the most dangerous special clause of all and can potentially cost a business hundreds of thousands of dollars.

A typical escalating MAC clause reads as follows:

> … subject to annual review, the volume commitment of this contract will be increased in each successive contract year, based on 90% of the customer's actual volume in the previous contract year…

For example, a large East Coast manufacturer signed a $1 million, 3-year, long-distance contract with its carrier. The company ignored the escalating MAC clause when it signed the contract. The first year was a prosperous year for the business, and its phone bills showed it. The company's total long-distance volume in the first year was $1.5 million. In year two, the carrier escalated the MAC from $1 million to $1.35 million, which was 90% of the actual volume of the first year. In the second year, the company's call volume dropped back down to the normal level of $1 million, and the carrier billed the company a shortfall penalty of $350,000. This customer would have saved a lot of dollars and headaches had it eliminated the escalating MAC clause during the initial negotiations.

Carriers do not want to become adversarial with their customers, so they may not require the customer to pay the full amount of the shortfall. But they will definitely use the shortfall situation to their advantage by forcing the customer to sign a second contract that might have much higher rates.

One way to avoid a MAC increase is to move all excess call volume to another carrier, as long as no exclusivity clause prevents the customer from moving traffic to another carrier.

Discount cap   This is the most ridiculous special clause because it actually punishes the customer for doing more business with the carrier! A typical discount cap clause reads as follows:

> Customer will receive a 30% discount on all long-distance usage charges. Discounts will only apply to *the first* $75,000 of long-distance usage per month.

In other words, the customer's discounts are capped off at $75,000. Any additional usage with the carrier will not be discounted. Many high-growth businesses have been burned by this clause. In the rare case that this clause is not accompanied by an exclusivity clause, the customer should simply move the excess traffic to another carrier. The best defense, of course, is to ensure that this clause is not included in the contract.

Automatic renewal   This clause causes the contract to *automatically renew* for another term at the end of the original term. Here is a typical automatic renewal clause:

> Upon the expiration of the term, this agreement will automatically renew for a period equal to the original term commitment, unless the customer provides written notice of customer's intent to terminate this agreement at least thirty (30) days prior to the termination of this agreement.

A customer at the end of a 3-year contract automatically begins a new 3-year term agreement with the carrier, unless the customer sends a cancellation letter to the carrier. The bad news with this clause is that telecom prices are always rising. (Rate increases are explained in Chapter 11.) To get lower pricing, a customer must forsake the old service offerings and sign a new contract with the carrier. If the old contract autorenews, then the customer will be forced to pay tomorrow the high prices of yesterday. The clause is not designed for carriers to lock customers into high pricing; it is simply a tool used by the carrier to get a customer's attention at the end of a term agreement. One surefire method of avoiding an automatic renewal is to send a cancellation letter early, even in the first few months of the initial term. Save a copy of the letter, as it may come in handy at the end of the term.

Special clauses that favor customers

The following are some of the most commonly encountered clauses included in a telecom contract to protect a customer's interests.

Business downturn   If a customer's level of business activity takes a downturn, his telephone calling volume will probably shrink. A company that loses a major government contract, for example, loses revenue and lays off employees. The company's phone bills will then be smaller because fewer people are talking on the phone. The business downturn clause protects a customer from shortfall charges. With this clause, the carrier will downgrade the customer's volume-commitment level, but require the customer to keep the remaining volume with the existing carrier.

Business divestiture   If a corporation spins off one of its subsidiaries, the daughter company will then be responsible for its own telecom services, and it will set up its own accounts with its own vendors. This volume will no longer show up on the parent company's phone bills and, therefore, will not contribute to the parent company's volume commitment. This may cause the parent company to miss its minimum annual commitment. The business divestiture clause protects a customer from shortfall penalties created in this situation. Of course, carriers are reluctant to include this clause, but with enough leverage, a business should be able to have it included in the contract.

90-day cancellation   Often called the 90-day out, this clause allows a customer to cancel the contract for any reason within the first 90 days. Here is an example of a 90-day out clause:

> ... if customer is dissatisfied with the provision of service within the first three billing periods ... customer may terminate this agreement without liability, and MCI will pay all costs of disconnection and reconnection to customer's prior carrier... (MCI Frame Relay Service Agreement Contract).

This clause is designed to limit the risk of trying out a new carrier. It will not be included in a contract for an existing customer, because carriers assume the customer has been satisfied thus far.

Technology upgrade   If a new telecom technology makes the existing service obsolete, the carrier will agree to adjust the volume commitment. The customer will not be subject to early termination fees as long as the new technology is procured from the original carrier.

For example, a Los Angeles news company had a $5,000 per month volume commitment with its long-distance carrier. The business spent $10,000 a month and found that $6,000 of it was intracompany calling. The company rerouted these voice calls across its existing frame-relay network and cut its long-distance phone bill to $4,000. Because of the technology upgrade clause, the carrier did not require the company to pay shortfall penalties.

Quality clause   If the telecom services do not work as promised, the customer may cancel the contract and switch to another provider without liability. The customer is normally required to allow the carrier 30 days to correct the problem first. Because telecom services are so critical to a business, some businesses switch carriers in troubled times even without having this clause in their contract. They may or may not be pursued by the carrier.

Service-level agreement   This clause clarifies the minimum quality level that the carrier guarantees to make available to the customer. The service-level agreement (SLA) is normally for high-tech data services such as frame relay and ATM. Two of the most important metrics covered in an SLA inculde the following:

- Percentage of time the network is available to the customer (usually 99.0% or higher);

- The carrier's maximum amount of response time prior to correcting any network problems.

Customers use SLAs to hold their carrier liable in the event that the carrier's network performs insufficiently. This clause should also spell out carrier penalties in the event that the carrier cannot provide the agreed-upon level of service.

Competitive offer   This is one of the most customer-friendly clauses around. A competitive-offer clause states that the incumbent carrier "… will match the rates of a legitimate offer from a competitive carrier…" The clause also states that the customer may terminate the agreement without liability if the current carrier cannot match the competitive offer. Because it gives too much power to the customer, most of the major carriers will not include this clause.

For example, McLeod USA has used this clause to get its customers to sign 5-year term agreements. No other carrier has been as successful in persuading small- and medium-size businesses to enter into long-term contracts. It can only be done when customers are assured they have the most competitive pricing in the industry.

Addendums   The special clauses mentioned thus far are standard clauses, but a business may have other concerns not addressed by these clauses. Such a business can write its own special clause and ask the carrier to include it as an addendum to the contract. The carrier's sales team will take the addendum up their chain of command for approval by high-level management, which will probably reword the addendum. Once both parties agree on the addendum's language, it can be included as part of the contract. Once again, if a customer has significant leverage in the negotiations, its addendum has a greater chance of being approved by the carrier.

Basic contract negotiation strategy

Contract negotiation normally begins with the customer handing over his telephone bills to a telephone company sales representative who later returns with a proposal that compares the customer's current costs to the telephone company's latest offering. The new proposal shows monthly and annual savings. The customer and sales representative then enter into a period of negotiation, when the telephone company's offer will be fine-tuned to meet the customer's requirements. Once the two parties agree on the offer, the sales representative will present a contract to the customer. The customer signs the contract, the phone company provides the services, the customer receives a bill each month, and everybody is happy. In the real world, however, telecom contract negotiation is never this smooth.

At best, contract negotiation is an annoying distraction from the core business; at worst, it is a nightmare. Account executives just want to close the sale, cash the commission check, and move on to the next target. Both parties have their own agendas, and they often wind up confused and frustrated by the process. There are some practical steps a customer can take to ensure that the negotiations are efficient and produce a favorable outcome.

Know what you want

First, the customer should know his own telecom environment. He should understand what telecom services his company uses and the monthly bill volume of those services. He should also know of any pending changes to the overall telecom environment.

Before talking with the carrier, customers should decide exactly what they want in their next telecom contract. What term commitment is the business comfortable with? Is the company's volume expected to increase, decrease, or remain level over this time period? What volume commitment should be made to the carrier? Will new services be added? After the customer figures out what he would like to see in the new contract, he can make a specific request of the carrier. A customer who is not specific is at the mercy of the carrier and will rarely receive the most favorable offer.

Know your carrier

Telephone companies handle their contracts two ways. Most contracts are standard templates, but some can be customized for a specific customer. Carriers employ two types of representatives: customer service representatives and account executives. Customers should know which type of representative they will be working with. Most importantly, the customer should know what elements of the contract are negotiable and nonnegotiable.

Customer service representatives are typically low-level employees, working in a call center. They normally can only offer standard contracts. Their hands are tied, and they have little room to negotiate. Account executives are usually highly paid outside sales representatives that are empowered to customize what is offered to the customer.

When negotiating with either type of representative, the customer's objective is always to get the needed telecom services at the lowest possible price. Carriers have the opposite objective; they want the customer to pay high rates, which increases their revenues and, ultimately, keeps shareholders happy.

Analyze the proposal

The highlight of any proposal is always the bottom line that shows monthly savings and annual savings. It is a good idea to double-check the calculations. "Accidental" spreadsheet errors may skew the numbers. Besides the savings, the other important aspects of the proposal are volume commitments, term commitments, pricing, and special clauses. If the offer is unacceptable, the carrier should write a new proposal documenting each change.

Pricing is often the most difficult issue for the customer and carrier to agree on, partly because customers do not know if they are being offered the carrier's best pricing. The most effective way to know if the pricing is fair is to compare the proposal to similar offers. Other carriers will be happy to present proposals, even if they know their chances of winning the business are slim. The customer will then know if the original carrier's offer is competitive. If the original carrier gets word that competitive carriers are submitting proposals, they usually sweeten their first offer.

A consultant can also be a valuable source of pricing information. If a consultant is hired to negotiate your contract, he will expect to be paid a percentage of the savings. You can save money by hiring the consultant on an hourly basis.

The contract

Once the final proposal is accepted, the carrier will offer a contract. Telecom sales representatives are trained to hand deliver the contract, verbally walk you through it, and ask for a signature. "This is the same info that we already discussed in the proposal. All the fine print is just a bunch of legal mumbo-jumbo. Go ahead and sign right here..."

Before signing, the customer should read the contract carefully. The contract must list everything the customer has negotiated, especially promotions, credits, and special clauses. This exercise may be a real eye-opener for the customer, because carriers may have thrown in additional conditions that were not previously discussed. Special clauses that are harmful to the customer often show up out of the blue, such as escalating MACs, traffic requirements, exclusivity, and discount caps.

Whether it is done intentionally or not, telecom carriers are notorious for performing "accidental" bait-and-switch maneuvers. Too many customers have negotiated promotions and aggressive pricing during the initial proposal phase only to find out later that these conditions were left out of the actual agreement. When the customer feels the sting and realizes what has happened, the original parties in the negotiations may be long gone. The original account executive has spent his commission check and probably moved into another profession. The new account team will have little sympathy for the customer, and the letter of the contract will rule.

The first phone bill

The final quality check in the process is to make sure that what was negotiated is actually showing up in the phone bills. If possible, schedule a meeting with your account executive to review the first month's bill. The bill might be easy to read, but this is a great opportunity to make the carrier prove that it has delivered exactly what it promised.

The first few bills of a new contract term are often inaccurate. Look for erroneous installation charges, missing discounts, and excessive fees. The first bill usually covers a partial month, so make sure charges are prorated accurately. With long-distance service, double-check the cost per minute. If the contract is with a new carrier, make your carrier aware of lines that are still billing with the other carrier. A savvy customer will require the carrier to issue invoice credits to make up for these costly glitches.

Local service

LECs only have three types of contracts: line charge, usage, and data services. LECs use contracts for line charges associated with non-POTS services such as Centrex, trunks, direct inward dialing (DID) lines, and intralata data circuits. LEC contracts for usage will cover local calling, intralata toll calling, or both. LEC data contracts will be included later with other data contracts.

Long distance

Long-distance carriers usually divide their customers into three categories: small and medium-sized businesses, large businesses, and national accounts. AT&T calls these three classes middle markets, commercial markets, and national accounts. AT&T has most often offered these markets Customnet, Uniplan, and One Net. The smaller market customers normally have simple one- or two-page contracts, while contracts used with larger businesses are 10 pages or longer. Other carriers categorize their customer base in a similar manner to AT&T. In general, however, the larger the carrier, the less flexible it will be with its contracts. Long-distance carriers providing other services, such as data circuits, will lump all of these commitments into a single contract.

Data

The data-network marketplace has been far less competitive than the long-distance marketplace. Consequently, contracts have remained simple and straightforward. Although the services are often higher dollar and are certainly higher tech, data-service contracts remain rather simple.

Wireless

Local, long-distance, and data services are all offered by large national carriers such as WorldCom, AT&T, and SBC Communications. The wireless industry has fewer giants and is comprised of smaller regional companies. Consequently, there is little consistency with wireless contracts. Each carrier has its own contracts, and the terms and conditions of the contract vary greatly from market to market and between individual service providers. The good news is that wireless-service pricing and contracts are simple and straightforward and contain few hidden surprises.

25

The request for proposal process

What is a request for proposal?

A request for proposal (RFP) is a document sent to telecom carriers from businesses seeking proposals from those carriers. Each RFP is unique, because each customer's telecom environment is unique. The RFP document spells out specific details about the information the business wants the carrier to provide, especially the technical data. Sending out an RFP is usually the first step in procuring high-level telecom services, such as data-network installations. The RFP is sent out to multiple carriers and is essentially an invitation to a bidding war.

What is the purpose of an RFP?

The main purpose of an RFP is to solicit proposals from phone companies. The average business signs 2-year telecom contracts, and at the end of that time, the business may have lost touch with telecom market trends, products, services, and, most importantly, pricing. It has no idea what a good deal in today's market is until it reads carriers' responses to the RFP.

After evaluating two or three RFP responses, the business will have a clear understanding of the carrier service offerings and what pricing is available. Gathering market data is, therefore, one of the key advantages of using an RFP.

Some customers require the telecom supplier to include the RFP as part of the final agreement. Without the RFP, the carrier's contract is the agreement. Carrier contracts are written by telephone company attorneys; they protect the interests of the carrier, not the customer. A well-written RFP, however, protects the interests of the customer, and, at the same time, sends strong signals to the carrier that the customer is in control of the relationship.

The main disadvantage of using an RFP is that the customer invests considerable time writing the RFP, evaluating RFP responses, and meeting with prospective carriers. This can be a time-consuming process. On the other hand, businesses can avoid the RFP process altogether if they are satisfied with their current carrier and are willing to allow the current carrier to provide the needed services.

What is the RFP process?

The RFP process consists of five phases:

1. The RFP is released to phone companies.

2. Phone companies question and clarify the RFP.

3. Phone companies submit proposals.

4. Customer evaluates proposals.

5. Customer selects the winning proposal.

What is specified in the RFP?

The core data in an RFP are descriptions of telecom services that the business has up for bid. These may be current services or future services that the business plans to add. RFPs can be used to procure local, long-distance, data and wireless services, network design, network installation, or any other imaginable telecom project.

An RFP explains the customer's expectations for customer service, service ordering, trouble reporting, billing, resolution of service outages, and SLAs. Numerous other issues can be included in the RFP. The whole idea behind an RFP is that the customer manages the procurement process, not the carrier.

A key feature of the RFP is its scalability. The scope of services covered by the RFP can be increased or decreased. A business may use an RFP to procure 25 cell phones, while another business may use an RFP to procure all of its telecom services, including local, long distance, data, and wireless.

For high-tech services, such as frame relay and ATM, the RFP should give the customer's specific technical requirements for these services. The RFP can be customized to include a large or small amount of technical data.

Who should use an RFP?

RFPs are normally only used by very large companies with complex, expensive telecom services. Smaller companies tend to solicit proposals informally, and they often do not have the time or manpower to devote to the RFP process. Larger companies have telecom departments, so they have the manpower to facilitate the RFP process. Bigger companies have more internal accountability and organizational layers, so the RFP also tends to appease these people inside the company. Many government agencies, for example, are required to secure multiple bids before entering into a new contract for services. Organizations that use RFPs tend to protect their interests more than businesses that shop informally for telecom services.

How does a phone company respond to an RFP?

After a phone company receives an RFP, a team of salespeople begins writing the response. The sales team consists of pricing experts, technology experts and possibly, on large accounts, regulatory experts. Their written response will follow the general outline of the RFP.

Carriers may try to gain a competitive edge by asking the customer questions to clarify vague parts of the RFP. Customers usually level the playing field by requiring that all questions be put in writing. Then, the customer sends each carrier a copy of the question and the customer's answer.

The RFP process is designed to be an objective avenue for buying telecom services. None of the carriers should gain an advantage over another bidder. No carrier knows which way the customer is leaning. The objectivity of the RFP process should keep carriers honest. Because carriers know they only have one chance to win the business, they will be more likely to give their best and final offer in the very beginning.

Appendix 25A
An RFP from the Acme Corporation for the provision of telecommunication services

The following is a sample RFP for a fictional business. The home office for Acme Corporation is in Memphis, but the company also has five separate field offices. The Memphis location operates as the hub for the company's WAN. The outer locations share data with the home office via frame relay connections. Each location has 5 to 10 local lines, with the exception of the Memphis office, which uses 35 local lines. Most of the corporation's long-distance calling originates at the home office, which also receives the inbound 800 long distance. The outside sales force of 50 employees actively uses calling cards, mobile phones, and a few pagers.

The Acme telecom manager is tired of dealing with multiple carriers and is looking for a single carrier to handle all of its telecom services. Numerous carriers boast that they can be a one-stop shop but Acme is reluctant to move everything to a single company at once. Instead, Acme's objective is to use a single carrier for frame relay, outbound long distance, inbound long distance, and calling cards. If the new carrier can handle these services, the lower-level services, such as local lines, wireless telephones, and pagers, will be moved to the new carrier.

Acme will send the following to AT&T, MCI WorldCom, Sprint, and Qwest. Acme is a fictional company, but there are about 1 million businesses in America that fit its profile. This sample RFP is scalable and can be reworked and customized to fit the needs of almost any business. Using an RFP like the following sample allows your business to manage the telecom buying process efficiently while maintaining control at each stage of the game.

Table of contents

Section 2: Proposal instructions

2.1   Project schedule

2.2   Preproposal conference

2.3   Receipt of proposals

2.4   Interpretation of documents

2.5   Submission of proposals

2.6   Acceptance or rejection of proposals

2.7   Evaluation criteria

Section 3: Descriptions and specifications

3.1   Network security

3.2   Management and operations

3.3   Implementation

Section 4: Carrier specifications

4.1   Your company

4.2   Customer service

4.3   Volume and term commitments

4.4   Long-distance service

4.5   Toll-free service

4.6   Calling-card service

4.7   Frame relay service

4.8   Staff-member services

Section 1: General scope

1.1 Project description

The Acme Corporation, hereinafter referred to as "Acme," is seeking a firm(s) to provide telecommunication services on a nonexclusive basis. This request for proposal (RFP) is an invitation for qualified firms to submit proposals to furnish, install, operate, and maintain telecommunication services for Acme. The term of the agreement shall be for a period of two (2) years. The specific areas of telecommunication services include long-distance, toll-free, calling cards, and frame relay.

1.2 Overview of Acme

Acme is a manufacturer of light industrial products. Acme, headquartered in Memphis, TN, currently supports five remote offices nationwide (see Appendix 25C).

1.3 Project scope

In an effort to consolidate services and thus provide increased cost savings to Acme, we are requesting that a firm(s) provide proposals to provide such stated telecommunication services to Acme.

Acme has an existing service agreement in place for both long-distance voice and frame relay/WAN data services. Acme's current telecommunication service agreement is coming up for renewal during the second quarter of 2001 and, as a result, we are evaluating the market for competitive bids on such stated services, via this RFP.

After receiving all qualified proposals, Acme will evaluate and then provide feedback to each company on the status of their proposed services.

Section 2: Proposal instructions

2.1 Project schedule

The schedule for this project is as follows:

Request for proposals released	December 30, 2001
Preproposal conference	January 8, 2002
Cutoff for questions	January 15, 2002
Proposals are due	February 5, 2002

2.2 Preproposal conference

A. A prepproposal conference will be held at Acme's corporate head-quarters. Questions submitted in writing by prospective carriers prior to the preproposal conference will be answered at that time or as promptly as possible thereafter. The fax number for submitting all questions is (901) 821-XXXX.

B. The following lists Acme's primary contact for this project:

Ms. Jane Doe
Director of Telecommunications
Acme Corporation
100 Main Street
Memphis, TN XXXXX
(901) 821-XXXX
(901) 821-XXXX (fax)
janedoe@acme.com

C. All inquiries concerning this RFP must be submitted in writing, via e-mail, or fax to the above location only. No other form of inquiry will be accepted.

2.3 Receipt of proposals

A. Sealed proposals, including three (3) sets of the Proposal Document for the Provision of Telecommunication Services, along with all supporting documentation will be accepted by Acme until 5 P.M. EST, on Monday, February 7, 2000.

B. Proposals may be mailed or hand-delivered to the address above.

2.4 Interpretation of documents

A. Questions concerning this document, contract terms and conditions, or additional information to prepare a response shall be submitted to Acme. Any interpretation of proposed documents will be made only by written addendum, and a copy of such addendum will be mailed and delivered to each carrier.

B. Failure to challenge or examine any work sites, system requirements, specifications, and instructions will be at the carrier's risk.

It is the responsibility of the carrier to inquire about and clarify any aspect of the RFP that is not understood. This RFP, all attachments, and the carrier's proposal, shall become a part of the final agreement.

2.5 Submission of proposals

A. This entire RFP document is to be returned with all supporting documents.

B. The proposal response must be formatted in the same order as outlined in this RFP.

C. No oral, fax, or telephone proposals will be accepted.

D. No carrier shall be prime respondent in more than one proposal. Collusion among carriers or the submission of more than one proposal under different names by any firm or individual shall be cause for rejection of all such proposals without consideration.

E. To submit a qualified proposal, the carrier must submit a proposal for all of the services, as outlined in this specification included herein. If all are not included, the proposal may be disqualified.

F. Pricing shall be considered to include all costs to provide the services required. Quoted prices are considered to be firm for a period of ninety (90) days from the date the proposal response is submitted.

G. All attachments shall reference the section and page number of item being discussed.

H. Carriers will provide a detailed installation schedule within the proposal. The selected carrier shall be required to enter into a contractual agreement based on this timetable.

I. The carrier may be required to submit a copy of regulatory certificates and tariffs authorizing the carrier to provide all services being proposed.

2.6 Acceptance or rejection of proposals

Acme's selection of a proposal will be based on an evaluation of the carrier as being both responsive and responsible, and as being the most advantageous to Acme. Acme reserves the right to waive any or all defects in form for any proposal. Acme reserves the right to reject any or all proposals. The selection of a proposal shall effect an agreement between Acme and the selected carrier.

2.7 Evaluation criteria

The following criteria will be used in the evaluation of each carrier and the proposed systems.

A. Financial offer: The costs for services, as well as any ancillary equipment, must be competitively priced. The ongoing operation of these services must also be competitively priced to include potential moves, adds, and changes that may be performed during the term of the agreement.

B. Technical proposal: Each carrier will be measured on its ability to respond to this proposal request in a concise, direct manner. The response must address items as directed, to include the following:

1. Reliability factors: System safeguards, real-time monitoring, service failure in terms of redundancy, built-in diagnostics, and common control failure protection.

2. System design: How well the services are expected to function, plus the technological aspects of interfacing with Acme's equipment, taking into consideration the multiple locations of the carrier's point of presence versus Acme's multiple locations.

C. Financial stability: Stability of the carrier will be reflected as follows:

1. Financial stability will be reflected by the carrier's financial statements.

2. Carrier size and credibility; in terms of professionalism utilized in design, engineering expertise, installation capability, training, and analytical support.

3. Separate ratings will be given for prime and major subcontractors.

D. Business organization and experience: This includes the overall ability of the carrier and its industry record in delivering substantial long-term support in providing telecommunication services and its ability to respond in the event of a disaster or malfunction.

Section 3: Descriptions and specifications

3.1 Network security

The carrier shall provide procedures to detect, minimize, and eliminate the possibility of a network security breach. This shall include network breach, toll fraud, and 976- and 900-number calling. Carrier must state exact dollar amount limits of fraud-related liability.

3.2 Management and operations

A. The carrier shall be fully responsible for the management and operation of its services, including the following:

1. Project administration: The carrier shall be responsible for the management and administration of the contract, including resources to perform the functions listed below. The carrier's project administration staff shall do the following:

a. Interface with Acme on issues related to the contract and its administration;

b. Oversee the overall management and operation of services provided under the contract;

c. Interface with Acme on issues related to operational support and implementation;

    d. Coordinate with Acme for all customer organizations, subcontractors, and other service providers during the implementation of services;

    e. Serve as a single point of contact to interface with Acme on issues related to trouble reporting and resolution;

    f. Notify Acme within one (1) hour of a major alarm or outage impacting services at Acme.

2. Project management: All resources assigned to fulfill the contract management and administrative functions shall be accessible to Acme twenty-four (24) hours a day, seven (7) days a week by telephone or pager. A list of all points of contact shall be provided. Any changes to the list shall be provided to Acme within seven (7) days of the effective date of change.

    a. Service ordering: The carrier's service order system should be explained in its response and shall incorporate the following functions:

        i. Order initiation: This includes service orders to initiate, add, change, move, or disconnect service and service features.

        ii. Service order tracking: The carrier shall provide a means for Acme to verify the status of service orders from initiation to completion.

        iii. Service order completion and acknowledgement: The carrier shall complete all testing and verification before delivering service to Acme.

    b. Operational support: In managing the overall operations for providing network and long-distance services, the carrier shall be responsible for the following activities:

        i. Inventory: Maintain an updated inventory of all circuits, equipment, services, and features of all services under this contract, and send an updated inventory report to Acme on a quarterly basis.

ii. Moves/adds/changes: Provide Acme with the capability for moves, adds, and changes of lines, services, and features through its service-ordering process.

c. Billing procedures: The carrier shall bill in arrears on a monthly basis and provide a single consolidated invoice for all services. Acme shall validate and pay the invoice within thirty (30) days of receipt.

i. The carrier shall provide a sample bill with its proposal.

ii. Invoice requirements: Complete invoices shall be provided via paper and electronic format.

iii. Each invoice shall contain all pricing components in sufficient detail necessary to reconcile charges with completed service orders or actual usage. Acme shall not pay any charges that are not specifically stated in the delivery order, including any new fees that the carrier charges other customers.

iv. Invoice preparation: The carrier shall prepare all invoices using location codes, in order for Acme to bill each Acme location.

v. Centralized billing: The carrier shall submit all invoices to the address below:

Accounts Payable
Acme Corporation
100 Main Street
Memphis, TN XXXXX

vi. Invoice data retention: All original invoices and related records shall be maintained by the carrier for the length of the agreement plus two (2) years. This data shall be available to Acme auditors within ten (10) days of any request.

vii. Service outage credits: The carrier shall credit Acme within two (2) billing cycles after a service outage. Credits shall be processed through the billing system and appear on the invoice.

viii. Billing disputes: Acme requires evidence that a service order has been completed and charges are priced correctly or it may dispute the charges. The carrier shall explain its dispute resolution methodology. The carrier shall provide a toll-free number and a single point of contact for dispute inquiries.

ix. The carrier shall resolve all billing disputes within thirty (30) days of notification by Acme.

d. Trouble handling: The carrier shall propose trouble handling procedures that include the following functions:

i. Trouble reporting: The carrier shall provide a single point of contact for trouble handling that is available twenty-four (24) hours a day, seven (7) days a week. Trouble reports should be received by toll-free number, e-mail, or fax. Records of all trouble calls shall be maintained for the duration of the agreement.

ii. Escalation procedures: The carrier shall propose an escalation procedure, with appropriate time intervals, for each service category. Point-of-contact names, titles, telephone and fax numbers, and e-mail addresses shall be provided to Acme.

B. Service response

1. Major alarms shall be corrected within two (2) hours. Remote maintenance is required to begin within fifteen (15) minutes of a major alarm. A major alarm is defined as any one or combination of the following system problems:

a. Catastrophic failure of switching systems or transmission facilities;

b. Disruption of service to critical users or circuits.

2. A minor outage (service call) is defined as any problem that is not considered to be a major alarm as outlined above. Service calls, which are reported during standard business hours, must be responded to and cleared up within two (2) hours.

C. Liquidated damages

1. Acme shall be entitled to liquidated damages in the event any outage or disruption of service is not corrected within eight (8) hours after the carrier's receipt of a trouble call. Acme shall receive a credit on the next monthly billing cycle invoice in the amount of the liquidated damages.

2. If a major outage has occurred and is not rectified or reduced to a minor outage by the end of the eight (8)-hour period, the liquidated damages shall equal $1,000 per eight (8)-hour period thereafter, until the major outage is rectified or reduced to a minor outage, up to a maximum amount of $5,000 per occurrence.

3. If a minor outage has occurred and is not rectified by the end of the eight (8)-hour period, the liquidated damages shall equal $500 and shall continue to accrue at $500 per eight (8)-hour period thereafter until the minor outage is rectified, up to a maximum amount of $2,500 per occurrence.

D. Reporting requirements

1. The carrier shall provide a sample of all reports described in this section as part of the RFP response. The following monthly reports shall be provided to Acme:

   a. Service order status summary: This report will provide the status of service orders from initiation to completion.

   b. Trouble report summary: This report will provide a summary of all trouble reports from their initiation through resolution.

   c. Monthly traffic statistics by service and location code.

3.3 Implementation

A. The carrier shall be responsible for managing and facilitating the transition and implementation of services, to include cutover testing and planning. The implementation strategy should do the following:

1. Meet service delivery schedules as required by Acme. The carrier shall submit a detailed implementation schedule to Acme within ten (10) days of award of the contract.

2. Assure that the services, functions, and features provided at the demarcation point conform to established industry standards.

3. Ensure seamless operations to Acme. The continuity and quality of existing services at Acme shall be maintained until the implementation of new services is completed.

B. Cutover testing

1. The carrier shall conduct cutover testing for each service category. The carrier shall submit a sample cutover test plan with its proposal and a detailed, service-specific cutover test plan within thirty (30) days of contract award. The plan should include site preparation requirements and system configurations.

2. If performance problems are encountered during testing, the carrier shall work with Acme and/or other contractors to isolate and eliminate the problems.

C. Training: The carrier shall provide training to Acme staff covering service requests, order status, trouble reporting, escalation procedures, billing, and report analysis.

Section 4: Carrier specifications

4.1 Your company

A. Enclose a copy of the most recent annual report or financial statement.

B. Briefly describe your network in terms of capacity, circuit redundancy, and switch nodes. Include a listing of those cities where you currently have digital switches installed.

C. Please enclose a network map indicating those cities serviced by your network.

D. Detail what percentage of your network is purchased from other communications providers.

4.2 Customer service

A. Describe your customer service philosophy.

B. Describe your service level agreement guarantee policies.

C. Describe your company policy on service-outage problem escalation for each product you are recommending (enclose a copy of your policy).

D. Describe your standard policy for financial reimbursement due to service outages.

E. Describe your fraud protection systems.

F. Explain your customer billing process and the flexibility of your payment options.

G. Enclose and describe how to read a sample consolidated monthly bill for each service that you are recommending to Acme.

4.3 Volume and term commitments

A. Include in your proposal one (1)-, two (2)-, and three (3)-year term commitments.

B. Include in your proposal three (3) different volume commitment levels. Consider Acme's volume amount as illustrated in Appendix 25B.

C. Include a sample contract for all services with your response.

4.4 Long-distance service

A. Describe the functionality of your long-distance service offering.

B. Describe the process for obtaining assistance with the functions described in point A.

C. Complete the following sentence: When a staff member initiates a long-distance call, we bill in _____ increments after the first _____, with a ____ minimum.

D. List and explain any switching or one-time fees to convert to your services.

E. Include proposed volume commitments that consider the call volume listed in Appendix 25B.

F. List all switched and dedicated rates, including intralata, intrastate, interstate, and international.

G. List any monthly minimum charges associated with your service offerings.

H. List any special times during the week (i.e., nights or weekends) when we would receive discounted usage rates.

I. State the carrier line charge.

J. State the Universal Service Fund and whether it is a flat charge or a percentage.

K. Enclose samples of all available custom reports.

L. List call types (i.e., directory assistance) that will not count toward meeting the minimum annual volume commitment.

4.5 Toll-free service

A. Describe the functionality of your toll-free service offerings.

B. Describe the process for obtaining assistance using any of the functions described in point A.

C. Complete the following sentence: When a customer initiates a toll-free call, we bill in ___ increments after the first ___, with a ___ minimum.

D. List and explain any switching or one-time fees to convert to your services.

E. Include proposed volume commitments that consider the call volume listed in Appendix 25B.

F. List all switched and dedicated rates, including intralata, intrastate, interstate, and international.

G. List any monthly minimum charges.

H. List any features included in your toll-free routing service.

I. Enclose samples of all available custom reports.

J. List any types that do not count toward meeting the minimum annual volume commitment.

4.6 Calling-card service

A. Describe the functionality of your calling card product.

B. Describe the process for obtaining assistance for problems using a calling card.

C. Describe, in detail, the features of your calling-card product.

D. List the costs associated with each of the functions described in point A.

E. Include proposed volume commitments that consider the call volume listed in Appendix 25B.

F. List and explain any monthly minimums associated with your calling-card product.

G. Describe how to obtain replacement or additional cards.

H. List any call types that do not count toward meeting the minimum annual volume commitment.

I. NOTE: The carrier shall provide long-distance calling cards to all staff, as requested by Acme, at no additional charge.

4.7 Frame relay service

A. Describe the functionality of your frame relay service offerings.

B. In answering point A, please include a list of the various fractional T-1 service levels offered (i.e., 56 Kbps, 128 Kbps) all the way through full T-3 service.

C. In answering point A, please include a list of the various bandwidth packet distribution levels and price breaks.

D. List any charges that will not count toward meeting the minimum annual volume commitment.

E. Describe the circuit ordering and provisioning process. Include how long it takes to turn up a circuit once it has been ordered. Identify how long the process takes at each step (i.e., the ticket will stay in the order entry phase for two (2) days).

F. Describe, in detail, the process required to deliver a circuit into an Acme location. Include everything, from a facilities site survey, to the extension of the physical circuit into Acme's suite and termination on our telco plywood. Who extends your circuit into the customer premises? Also, describe the physical attributes of this type of circuit, including the cabling.

G. Describe, in detail, all volume discounts that would be made available to Acme via installation pools, etc., regarding frame relay service.

H. Describe, in detail, your approach to proactive circuit monitoring, as well as the troubleshooting process invoked with both minor and major WAN outages. Include a description of the entire circuit recovery process, beginning with the opening of a trouble ticket (by either Acme or your company) and ending with the closing of the ticket. Include a sample of your circuit outage detail and summary reports.

I. Include samples of all monthly, quarterly, and annual circuit usage reports.

J. List the recommended customer equipment required for frame relay circuit deployment, including the cost of leasing this equipment from your company.

4.8 Staff-member services

Describe any benefit packages you would be able to offer to Acme staff members, on a companywide basis. Describe the actual features, functions, and rate structures associated with each service being offered. Sample packaged services include long-distance, calling cards, Internet, or wireless services.

Appendix 25B
Acme's long-distance call volume

The figures below represent usage in a typical month. The figures are based on a 6-month average. Memphis has dedicated T-1 service. All other locations are switched.

LOCATION	INTRALATA	INTRASTATE	INTERSTATE	INTERNATIONAL
Outbound Direct-Dial Long Distance				
Memphis, TN	3,205	1,654	15,216	1,996
Alexandria, VA	1,850	3,542	2,548	364
Charlotte, NC	210	5,241	3,245	564
Dallas, TX	2,310	4,251	8,354	984
Syracuse, NY	0	954	1,254	1,245
W. Palm Beach, FL	4,210	4,215	7,215	889
Inbound Toll-Free Long Distance				
Memphis, TN	245	5,364	27,548	1,154
Alexandria, VA	325	1,235	1,287	0
Charlotte, NC	145	7,183	7,547	0
Dallas, TX	457	5,421	11,258	0
Syracuse, NY	0	0	0	0
W. Palm Beach, FL	567	2,155	12,541	0
Calling-Card Long Distance				
Memphis, TN	599	354	1,551	0

Appendix 25C
List of Acme's sites' frame relay information

Location	NPA/NXX	Frame Relay Port Speed
Memphis, TN	901-821	3.01 Mbps
Alexandria, VA	703-739	256 Kbps
Charlotte, NC	704-529	256 Kbps
Dallas, TX	972-490	768 Kbps
Syracuse, NY	315-432	256 Kbps
W. Palm Beach, FL	561-687	256 Kbps

Glossary

access the connection between the end user and the network

airtime charges for usage on a mobile phone

association discount an additional discount, usually 5%, that is added to an account because the customer is a member of an association that has an alliance with the carrier

ATM asynchronous transfer mode, a high-speed packet-switching service

backbone high-capacity lines in a carrier's network that link the carrier's central offices

bandwidth the amount of data that can be passed along a communications path

central office the central building in a carrier's area where all the lines converge. The central office equipment switches the calls.

circuit a generic term for any telecommunications physical connection

circuit switching a method of data networking whereby the sending computer connects to the receiving computer using a dial-up method. The circuit remains in use until the computers are finished communicating and they hang up. ISDN is a circuit-switching technology.

class of service the type of local lines a customer has. The lines are either residential or business; lines, trunks or Centrex; and flat, message, or measured rate.

contingency plan a back-up plan to reroute data traffic in the event that the primary data network fails

contract value the financial value of a customer to the carrier. Contract value is calculated as months left in contract × monthly commitment.

convergence the merging of voice and data across the same data network

cramming the illegal practice of adding unauthorized monthly charges to a customer's account

dedicated a line between two points in a data network. The line is always available.

dial around the practice of using a different long distance carrier by manually dialing the new carrier's code

direct dial an outbound long distance call dialed by the user without operator assistance

directory services yellow pages, white pages, and directory assistance

digital subscriber line a new high-bandwidth technology that allows residences and small businesses to make voice calls and connect to the Internet across ordinary copper wires

exchange used either to describe a telephone company's local calling area or the local access and transport area (LATA)

extranet a company data network that allows outside customers and suppliers to share the network

frame relay a high-speed packet-switching service

home area the local market for a mobile phone customer

hunting/rollover   service provided by the local carrier that transfers calls to the next line if the first line is busy. If the last number is busy, the call will rollover back to the first number

inbound   long-distance call to a toll-free number

interexchange   a call between local exchanges or LATAs; long-distance call

interlata   a call between LATAs; a long-distance call

international   a call between countries

Internet   a huge global network composed of multiple smaller networks that use the same TCP/IP protocol

interstate   a call between states

intralata   a call within a LATA

Intranet   an internal company network using Internet standards and the TCP/IP protocol

intrastate   a call within the state

integrated services digital network   ISDN, a digital technology that allows users to send voice, data, and video

last mile   the concept that telecom networks are hampered by the slowest link, the copper-wire connection between an end user and the carrier

local loop   the telephone line in a data circuit from the local telephone company's central office and the end user

media   telecommunications signals travel on media. Physical media may be copper wire, coaxial cable, or fiber-optic lines. Broadcast media may be radio waves or microwaves.

multiplexing   a technique that allows multiple devices to share the same phone line

off-net   a location in a virtual private network that is not connected to the others by a dedicated connection

on-net   a location in a virtual private network that is connected to the other locations by a dedicated connection, such as a T-1

outbound   a direct-dialed long-distance call

packet switching   a data networking technology that breaks the data stream into individual packets of data prior to sending them out into the network. Frame relay and ATM are packet-switching technologies.

PIC freeze   a PIC freeze is implemented by the local telephone company. It prevents a customer's long-distance carrier from being switched without the customer's written authorization.

point of presence   a long-distance carrier's closest central office in a local market

rerate credit   a refund credit given to a customer to offset higher rates paid due to a carrier's mistake

"ring to" number   the POTS line an 800 number terminates into; also called the "pointed to" number

roaming   using a mobile phone while outside the caller's own home area

shortfall   failing to meet the volume commitment

slamming   the illegal practice of changing a customer's long-distance carrier without authorization

special clauses   clauses added to a contract that modify the standard terms and conditions of the contract

supercarrier   a carrier capable of providing all of a customer's voice and data telecommunications services

surcharge   the setup charge for a calling-card call

switched service   making calls that are switched at the local carrier's central office; the opposite of dedicated service

T-1   a technology with 24 distinct communication paths that is capable of transmitting 1.544 Mbps. T-1 service is used for access and point-to-point data connections.

tariff   a document filed with state regulatory bodies or the FCC. Tariffs list details and pricing of a telecommunications service.

telco   telephone company, usually used for local telephone companies

term commitment   the length of time a customer agrees to stay with a carrier

tie line   a dedicated line between two customer locations, usually used for voice traffic

volume commitment   the amount of money a customer agrees to spend with a carrier

List of acronyms

ATM   asynchronous transfer mode

BOC   Bell Operating Company

BRI   basic rate interface

CAP   competitive access provider

CIC   carrier identification code

CIR   committed information rate

CLEC   competitive local exchange carrier

CPE   customer premise equipment

CPM   cost per minute

CSR   customer service record

DID   direct inward dialing

DSL   digital subscriber line

DWDM   dense wave division multiplexing

ESN   electronic serial number

EUCL   end user common line charge

FCC   Federal Communications Commission

ILEC   independent local exchange carrier, or incumbent local exchange carrier

ISDN   integrated services digital network

ISP   Internet service provider

IXC   interexchange carrier

LAN   local-area network

LATA   local access and transport area

LDDS   Long Distance Discount Savers

LEC   local exchange carrier

LEOS   low earth-orbiting satellites

LOA   letter of agency

LPIC   intralata PIC

MAC   minimum annual commitment

MAN   metropolitan area network

MCI   Microwave Communications, Inc.

MIN   mobile identification number

MSA   metropolitan statistical area

NAM   numeric assignment mode

OPX   off-premise extension

OSP   operator services provider

PBX   private branch exchange

PCS   personal communications service

PIC   primary interexchange carrier, or presubscribed interexchange carrier

PICC   primary interexchange carrier charge

POTS   plain old telephone service

PPU   pay per use

PRI   primary rate interface

PRP   pager replacement program

PUC   public utility commission

PVC   permanent virtual circuit

RBOC   Regional Bell Operating Company

RCF   remote call forwarding

RESPORG   responsible organization

RFP   request for proposal

RSA   rural service area

SLA   service level agreement

SONET   synchronous optical network

USOC   universal service order codes

VoFR   voice over frame relay

VoIP   voice over Internet

VPN   virtual private network

WAN   wide-area network

WAP   wireless access protocol

WATS   wide-area telephone service

Selected bibliography

Brosnan, M., J. Messina, and E. Block, *Telecommunications Expense Management,* New York: Telecom Books, 1999.

Dodd, A. Z., *The Essential Guide to Telecommunications,* Upper Saddle River, NJ: Prentice Hall PTR,1997.

Oslin, G. P., *The Story of Telecommunications,* Macon, GA: Mercer University Press, 1992.

About the author

S. C. Strother has worked in telecommunications since the early 1990s. He began as an entrepreneur in long-distance service and privately owned payphones and went on to work for a regional carrier, selling and implementing contracts with new business customers. As a consultant, Mr. Strother represented more than 200 corporate customers in their negotiations with the telephone companies. He audited their phone bills and returned more than $1 million back to them. In addition to his research and writing, Mr. Strother has developed telecommunications training and has taught hundreds of auditors in the art of telecommunications cost management. He audits telephone bills and serves as a consultant to businesses and corporations.

Mr. Strother holds an M.B.A. and is currently working toward a Ph.D. at the University of Louisville.

Index

Understanding Networking Technology: Concepts, Terms, and Trends, Second Edition, Mark Norris

Videoconferencing and Videotelephony: Technology and Standards, Second Edition, Richard Schaphorst

Visual Telephony, Edward A. Daly and Kathleen J. Hansell

Wide-Area Data Network Performance Engineering, Robert G. Cole and Ravi Ramaswamy

Winning Telco Customers Using Marketing Databases, Rob Mattison

World-Class Telecommunications Service Development, Ellen P. Ward

For further information on these and other Artech House titles, including previously considered out-of-print books now available through our In-Print-Forever® (IPF®) program, contact:

Artech House
685 Canton Street
Norwood, MA 02062
Phone: 781-769-9750
Fax: 781-769-6334
e-mail: artech@artechhouse.com

Artech House
46 Gillingham Street
London SW1V 1AH UK
Phone: +44 (0)20 7596-8750
Fax: +44 (0)20 7630-0166
e-mail: artech-uk@artechhouse.com

Find us on the World Wide Web at:
www.artechhouse.com